Emily K. da Silva Cunha Souto
Flávia Lins Silva

New directions for experimental lessons in science teaching

Emily K. da Silva Cunha Souto
Flávia Lins Silva

New directions for experimental lessons in science teaching

Myth, Mystery, Necessity?

ScienciaScripts

Cover image: www.ingimage.com

This book is a translation from the original published under ISBN 978-613-9-73755-0.

Publisher:
Sciencia Scripts
is a trademark of
Dodo Books Indian Ocean Ltd. and OmniScriptum S.R.L publishing group

120 High Road, East Finchley, London, N2 9ED, United Kingdom
Str. Armeneasca 28/1, office 1, Chisinau MD-2012, Republic of Moldova, Europe
Printed at: see last page
ISBN: 978-620-7-88192-5

DEDICATORY

I dedicate this work to God, because this research was only possible under his blessings, to my dear husband Ivanildo, for his companionable presence, for his encouragement and for the love that strengthens me every day, to my family, teachers and friends who have always believed in my work.

ACKNOWLEDGEMENTS

To God, for his blessings throughout this journey, for enlightening me and giving me the strength to carry on in difficult times.

To my husband Ivanildo, for his encouragement and understanding in times of stress, for understanding my absence and lack of time and, above all, for his love and encouragement.

To my parents Petrônio Everson and Elizabete Cunha, who conceived my life, for their encouragement in every decision made in my life.

To my siblings Priscila and João, who have always supported and helped me throughout this journey. Especially for their words of love and affection.

To my advisor Prof Dr Flávia Lins, for her competence, commitment, respect, dedication and, above all, for her patience and affection with which she guided this work.

To my fellow students who made this journey unforgettable, for their friendship and affection at all times.

The Papa Paulo VI Municipal Elementary School and the teachers where I carried out this research, for the opportunity and support they gave me to carry out this work.

To the 7° year-olds who took part in this research, for the marvellous teaching-learning moments.

To the professors of this examining board, Dr^{3} Marisa Apolinário and Dr^{3} Carina Scanoni, for their critical and detailed reading of this work.

To the teachers of the Biological Sciences Degree Course for their teachings.

My cousin Liliane, who was always willing to take time out of her day to read and exchange information.

My friends Ana Paula, Cristiane, Mislene and Estefania for their advice and encouragement during this final stage of the course, I know I can always count on you.

The girls in rooms 12 and 8 of the Women's University Residence, Cristiane, Laudenize, Simone, Adriana, Gilmara and Priscila for the conversations, laughs, advice and countless moments of joy.

And finally, to all my friends, especially my friend Laudenize Souto, who was always thrilled with my achievements, and who was by my side throughout the course, accompanying me through all the experiences of this research and helping me compose my work.

"To teach is not to transfer knowledge, but to create possibilities for its production or construction. Those who teach learn by teaching and those who learn teach by learning."

(Paulo Freire)

SUMMARY

Investigative experimental activities have become an extremely important strategy in the teaching-learning process in science education, as they allow concepts to be learnt and important skills to be developed. Based on these assumptions, this work presents an experience report with investigative experimental activities in a public school in Nova Floresta - PB. °The work was carried out with students from two primary school classes and aimed to evaluate the application of experimental lessons as a pedagogical resource for the teaching-learning process. In this study we sought to develop a methodology that, as well as characterising the students' thinking throughout the investigative activity, would allow us to relate each student's previous knowledge to their performance and strategies used during the activity, relating it to some everyday problems. The proposed experimental activities were designed taking into account three themes: Microbiology, Evolution and Plant Morphology and Anatomy. With this teaching method, the students were more interested and participative, and a significant change in their understanding of the content was noticeable. It is therefore clear that experimental lessons in science teaching are stimulating and fundamental for students to build their knowledge.

Keywords: Experimental lessons; Science teaching; Teaching and learning; Primary school.

SUMMARY

CHAPTER 1

INTRODUCTION

Over the years, science teaching has had different objectives based mainly on the changes in society at different times (PCN, 1998; ROSA and ROSA, 2012; ZÔMPERO and LABURÚ, 2011). Today, with constant scientific advances and discoveries, schools need to be constantly renewed, making it necessary to train students who are more critical and attentive to these transformations. According to Belilacqua and Silva (2007), the information produced is highly specialised and accelerated, and access to this information is dynamic and almost instantaneous. This makes it urgent to structure processes that aim to ensure that scientific and technological knowledge is appropriated by society as a whole, in order to equip it to form an opinion and take informed action (MALACARNI and STRIEDER, 2009). In this way, science teaching and experimental classes at school are of total importance and can have their results improved if used for learning scientific knowledge (MALACARNI and STRIEDER, 2009).

> New educational methodologies must link what is learnt in the classroom with what students experience in their daily lives. With regard to the teaching of natural sciences, it can be seen that, in general, students have faced difficulties in assimilating the contents of this area of knowledge. It is likely that these problems are due to the lack of practical activities in primary school science classes, as well as the lack of teacher training (BELILACQUA and SILVA, 2007, p. 2).

Experimentation is a fundamental activity in science teaching, with the potential to motivate students, encouraging reflection on proposed themes, stimulating active participation in the development of the lesson and contributing to the effective possibility of learning. According to Carvalho et al (2010), the topics taught require practical lessons, thus forming a scientific attitude, which is inextricably linked to the way knowledge is constructed. As well as helping to develop scientific knowledge, practical lessons allow students to learn how to objectively approach their world and how to develop solutions to complex problems.

As well as motivating students, experimental classes have a very important transformative role, as Possobom et al. (2003) state, as well as being a place for learning, the laboratory is a place for developing the student as a whole. Also according to Vasconcelos et al. (2002), in practical classes students develop procedural skills linked to the scientific process, such as observation, inference, measurement, communication, classification and prediction. From these, or concomitantly, integrated skills are developed: control of variables, operational definition, formulation of hypotheses, interpretation of data and conclusion.

> Procedural and integrated skills are closely linked to the objectives of science teaching, because they arouse curiosity and interest in nature, stimulate the habit of study and

observation, which are necessary conditions for improving logical thinking and developing inductive and deductive reasoning. In science lessons, therefore, students have to manipulate laboratory materials, observe, mix, measure temperatures, complete tables and calculate averages. Modern and creative scientific practice must therefore include a set of procedures that bring students closer to more rigorous and creative ways of working, more consistent with the way scientific knowledge is produced (VASCONCELOS et al.; 2002, p 3).

Experimental lessons should not be limited to manipulation and naming of glassware and reagents, but it is essential to guarantee space for reflection, development and construction of ideas (POSSOBOM et al.; 2003). Activities such as pipetting, calibrating instruments and preparing solutions, among others, contribute little to the student's relationship with society (BUENO and KOVALICZN 2009).

Carrying out investigative experimental activities in science teaching is an excellent tool for students to consolidate the content and build an affinity between theory and practice. In this way, such activities should be developed based on investigative questions that take into account the students' lives and that pose real and challenging problems with the aim of going beyond the manipulation of laboratory materials (BUENO and KOVALICZN 2009; CASTOLDI and POLINARSKI2009).

According to Bueno and Kovaliczn (2009), the experimental activity should provide conditions for students to raise and test their ideas and assumptions about the scientific phenomena that occur in their surroundings, so the teacher's role is to guide, mediate and advise on the process.

> Experimental activities should be seen as situations in which students learn to make conjectures and interact with their colleagues and the teacher, exposing their points of view, their assumptions and confronting mistakes and successes. In this way, laboratory experimentation helps students reach higher levels of cognition, which facilitates the learning of scientific concepts and their social purposes (BUENO and KOVALICZN 2009, p 3).

It can therefore be considered that investigative activities in which students seek an answer to the problem question help to achieve any suggested objective, since experimentation intensifies the understanding of concepts and leads to meaningful learning. But in order for the experimental activity to achieve its objectives, the teacher needs to have knowledge of the topics to be covered, the skills to carry out the experimental lessons in an engaging way, mainly including the student in the execution and reflection of the data, because it is in this investigative environment that the student learns. As Hodson (2005) apud Silva (2011) argues:

> The activities must involve the students cognitively, they must exchange ideas with each other and with the teacher, and they must lead to the construction of scientifically significant knowledge. Practical work becomes more relevant when teachers are clear about its purpose and when they plan activities so that these purposes are achieved. (HODSON 2005 apud SILVA 2011 p. 37)

Although experimental lessons are considered by teachers to be valuable tools in the teaching-learning process, they are almost absent from the classroom, occurring only a few times and with objectives that differ from, and even conflict with, the proposals that advocate them.

According to Malacame and Strieder (2009), these difficulties in carrying out experimental lessons are due to the reality of teacher training, which lacks reflection on science and its teaching, causing great insecurity about the development of scientific knowledge in the classroom; and resulting in little or no innovative work, limited in many cases to reading or carrying out exercises proposed by the textbook which, however well produced, contributes little to an attractive first contact for children with the dynamic world of science.Faced with so many doubts about how to deal with and take advantage of experimental education, this research aimed to improve the application of experimental science lessons at the Papa Paulo VI Municipal Primary School, with a view to building scientific knowledge for the student.

CHAPTER 2

BACKGROUND

It is well known that education is a nation's greatest legacy. However, the educational process is not a one-way street. Today, with technological advances and the innovations that the market offers us, it is up to the teacher to deepen the theoretical subjects discussed in class with practical examples. From experience, when theory is reinforced with didactically practical examples, it can be seen that adolescents' curiosity is sharpened, awakening criticality and a broad world view and the power to question.

The teacher actually serves as a bridge to broaden students' knowledge, especially when he or she encourages students to think critically, where questions asked in strategic ways have been at the heart of the teaching profession since the advent of teachers and students. A practical lesson followed by a good sequence of questions makes it possible to build solid mastery even of complex ideas, because it reveals and explains. Questioning is therefore the construction of steps, and if this structure is solid, it will enable the student to rise to any height, because by climbing these steps the student will no longer look at their feet, as they will be able to climb and develop and decompose ideas in a systematic way.

CHAPTER 3

OBJECTIVES

3.1 General Objective

- Evaluating the inclusion of experimental lessons as a pedagogical resource for the teaching-learning process in the 7th year of Primary School II at the Papa Paulo VI Nova Floresta Municipal School - PB.

3.2 Specific objectives

- To identify how experimental lessons can contribute to the reasoning processes in the formulation of relationships between theoretical content and educational practice in the stages of science knowledge production;
- Classify the instruments and resources that can contribute to the formulation of practical models that dimension science teaching.
- To provide support for better use of the lessons.
- Learn about possible materials and resources for working with students in science education.

CHAPTER 4

THEORETICAL BACKGROUND

4.1 . Traditional Lessons *versus* Modern Method - MYTH

During moments in the history of education, the role of teaching has become more important than learning and even today, some teachers have shown that traditionalist teaching permeates the school environment, where the transmission of historically accumulated knowledge is emphasised (RODRIGUES et al.; 2011). In this view of teaching, students are seen as mere listeners (passive subjects) who must memorise, repeat and obey; such students have a totally limited role and take little part in the development of knowledge; memorising is what must be done without any effective learning strategy.

It is clear that this type of teaching is even more evident in practical classes. You only have to enter a classroom to realise this: teachers try to demonstrate certain phenomena using only the blackboard and chalk, often without even the necessary discussion to explain it (FARIAS et al., 2009; PEDRANCINI et al., 2007). According to Leão (1999), the relationship that is established between teacher and pupils, when the former lectures and the latter take notes and memorise, is not conducive to learning; on the contrary, it makes it difficult or impossible for it to take place.

Due to these teaching concepts, the laboratory, which is often seen by students as a place that arouses curiosity and interest in science, has also been neglected. When it is used, it is only used to clarify content or simply to demonstrate experiments, which often have little to do with the students' reality. In this way, the simple fact of studying science using a traditional approach doesn't help to build the set of skills and abilities to organise new knowledge (PRIGOL, GIANNOTTI, 2008; VASCONCELOS et al, 2002).

> Traditional teaching proved to be inefficient when priority began to be given to teaching the "scientific method", and so a new way of teaching science emerged: teaching by rediscovery. In practice, this proved to be mistaken, but it had the merit of breaking with the tradition of teaching science by transmission-reception (CAMPOS and NIGRO, 1999, p.26).

It is no longer acceptable for science teaching to be limited to transmitting certain concepts to students. Nor is it acceptable to rediscover the scientific method, because just observing and recreating is not doing science, because even if students think they are living the scientific method, what they are doing is just making generalisations similar to those made by some scientists of the past (PRAIA et al, 2002). Learning in this way can confuse students as well as teachers because they are unclear about their role in the teaching-learning process. Science is more than studying unknown phenomena, so science teaching should provide students with the opportunity to develop skills that arouse concern about unknown facts, so that they can seek logical and reasonable explanations for the events they seek to understand (HOERNING, PEREIRA 2004). As Bizzo

(2008) confirms:

> "In this way, students will be able to develop critical attitudes, make judgements and decisions based on criteria that are as objective as possible, defensible and based on knowledge shared by a broad school community."

In this view, the teacher serves only as an organiser of teaching circumstances. Today, the teacher is the one who teaches in a way that encourages active teaching, in which the student is the subject of the action that allows them to move from initial forms of knowledge to critical awareness. The teacher must focus on the relationship between content and life experience, so that students feel they are in a single world and not in two separate worlds: school and life outside school (SUART et al, 2009; BUENO, KOVALICZN 2008).

The teacher must act in an innovative way, with an investigative eye based on a critical view of education, where the student is the focus of discussion. Methodologies of this kind make the world of the student in the classroom interesting, because today with technological advances the world outside the classroom is much more interesting. It's up to teachers to bring advances into the classroom and use them with practical and didactic models, where classes with just a blackboard and chalk will be replaced by attractive and curious classes (SÁ 2009; JÚNIOR, FERREIRA, HARTWING 2008).

In this sense, teaching science and biology means allowing students to actively participate in research, because as they appropriate scientific knowledge, they can prepare themselves for the world around them. To achieve this goal, biology teachers need to be able to include in the classroom not only the problematised content set out in the curriculum, but also the questions, contributions and uncertainties presented by the students. As the problem situations are investigated and recognised by the students, the obstacles are mediated by the teacher. As Silva Júnior and Barbosa (2008) confirm:

> Teachers of science and related subjects are charged with being continually up-to-date and in sync with all these scientific dynamics. However, what will determine student learning at all levels of education, to the detriment of memorised content that is forgotten after assessments, are the didactic forms that teachers in this area of knowledge will use (SILVA JÚNIOR, BARBOSA, 2008, p 14).

When science teaching is treated only as a conceptual change, it ends up excluding or cancelling out the student's power of investigation, so that they don't think, they don't investigate, they just superficially accept what they have been taught. In this way, students tend to generalise uncritically, come up with quick answers based on common sense evidence, and think in a linear way (CAMPOS and NIGRO 1999). It is therefore necessary to overcome this methodology of superficiality. Students should carry out experiments that bring them closer and closer to "doing science", so that they have opportunities to face real problems and seek solutions to them. To do this, they will initially use prior knowledge, but they will soon be moving on, using new ideas, overcoming evidence of common sense, becoming more critical and rigorous, formulating and comparing hypotheses in their various situations, so that they stop thinking about things in the world in a superficial way (SILVA, 2011). As Campos and Nigro (1999, p. 30) point out:

> "When teaching and learning science as an enquiry, the starting point is the problems that the students face - in this sense, the cognitive conflict is not an external imposition. The students themselves work with explanatory hypotheses and, as a result, they compare them all the time, test them, put them in situations of potential conflict, in order to validate them or not."

According to Moreira (1999), many teaching models are based on Jean Piaget's theory of cognitive development. It is based on the perspective that the human mind permanently tends to increase its degree of internal organisation and adaptation to the environment. When faced with new information, there are imbalances and consequent restructuring (accommodation) in order to build new assimilation schemes and reach a new equilibrium, guaranteeing a greater degree of cognitive development. In this way, teaching (or, in a broader sense, educating) means provoking imbalance in the child's organism (mind) so that, seeking to rebalance (major equilibration), it can restructure itself cognitively and learn (MOREIRA, 1999).Thus, we realise that science only advances with experimentation, and it is necessary to abandon traditional teaching models where transmission and mere repetition are the mainstays. This is the only way we can really move forward and encourage our students to investigate and experiment with new knowledge.

4.2 The importance of experimental activities in building knowledge - MYSTERY

We know that the importance of practical activities is unquestionable, but this teaching model has often been approached in a superficial and intuitive way, devaluing its cognitive aspects. Much discussed among researchers and teachers, experimental classes should mediate the teaching-learning relationship, since they stimulate student interest in the classroom, providing participation in subsequent activities, thus constituting a key aspect in the teaching and learning process (ARAÚJO E ABIB 2003, JÚNIOR, FERREIRA E HARTWING 2008, PEDRANCINI ET AL 2007).

According to Campos (1999), practical activities are classified into: (1) Practical Demonstrations: Activities carried out by the teacher, which the student watches without being able to intervene, (2) Illustrative Experiments: Activities which the student can carry out and which fulfil the same purposes as practical demonstrations, (3) Descriptive Experiments: Activities which the student carries out and which are not necessarily directed all the time by the teacher and (4) Investigative Experiments: Practical activities which require a great deal of activity from the student during their execution. They differ from the others in that they necessarily involve discussing ideas, drawing up explanatory hypotheses and carrying out experiments to test them.

Of these, the one that draws the most attention and will be prioritised in this research is investigative experiments, which have been considered by various researchers as a teaching and learning strategy that can help develop students' cognitive skills in such a way that their work is not limited to manipulation and observation. Students have the chance to take part in all phases of the investigation, such as suggesting their hypotheses for the problem, collecting data, analysing it and drawing conclusions based on the assumptions made, actively participating in the construction of knowledge.

Tamir (1977) argues that investigative experimental work has the following characteristics: (1) it should be a means of exploring students' ideas and developing their conceptual understanding; (2) it should be underpinned by a prior theoretical basis that informs and guides the analysis of the results; (3) it should be designed by the students to allow them greater control over their own learning, over their difficulties and to reflect on the reason for these activities in order to overcome them.

Investigative activities look for a problematising question that simultaneously arouses curiosity and guides the student's point of view on the important aspects of the fact being studied, making them raise their own assumptions and indicate possible solutions (HAMBUGER and LIMA 1988, ZÔMPERO and LABURÚ 2011). The problem to be investigated should be chosen taking into account the students' interest so that they feel motivated to solve it.

> It is therefore important for teachers to be clear that teaching natural sciences is not just about presenting scientific definitions, as in many textbooks, which are generally beyond the students' comprehension. Definitions are the end point of the teaching process, what students are supposed to understand and systematise during or at the end of their investigations (HAMBUGER and LIMA, 1988, p 13).

It should be remembered that many teachers tend to think that the fact of asking students open-ended questions is enough to guarantee the assumption of real problems. In fact, a real problem is one that causes a cognitive conflict that we can't come up with an immediate answer to, in other words, it's a situation that is different from what we understand and propose solutions to (ARAÚJO et al 2012, SOUZA 2007).

In this way, students may realise some contradictions and flaws in their way of thinking (CAMPOS and NIGRO 1999). These doubts will then become the starting point for the creation of ideas that help them seek explanations for previously unquestioned knowledge, causing dissatisfaction and helping them to feel the need to modify these ideas. As Campos and Nigro (1999) state, this cognitive conflict can suggest a path along which knowledge will be constructed.

An important aspect that can be noted is the need for investigative activities to provide students with a direct relationship with new information. In investigative activities, it is necessary to communicate the new knowledge obtained by the students. So the teacher must try to change not only the student's way of thinking, but also the methodology, so we don't just have to transform what we think, but also how we think. For Zômpero and Laburu, (2011, p 75):

> An important aspect that can be observed is the need for investigative activities to provide students with contact with new information. In investigative activities, it is necessary to communicate the new information obtained by the students.

It is necessary for the student to stop adopting the methodology of superficiality in order to employ a methodology that is closer to that of science. As Francisco Júnior, Ferreira and Hartwig (2008) state, old knowledge is overcome by new knowledge in an incessant and restless movement that, at the same time, respects the ideas brought by the students and assumes them as historical beings who "are being". The aim is always to reformulate the students' ideas, bringing them ever closer to the scientifically accepted concept.

Zuliani (2006) points to an investigative methodology, in which the student must play an active role in the learning process, in a procedure in which the student becomes familiar with the richness of the activity of constructing the knowledge under study, starting to see it as an open and creative activity. It also considers that the student should be effectively engaged in the process, and therefore it is essential that the proposed task is challenging and meaningful, and that it also allows choice and control over the activity itself.

> The use of didactic laboratories based on real, everyday student data, based on the process of investigation, carried out within the epistemological vision of science as a human activity, can become an important factor in the conceptual evolution of the learner. On the other hand, the search for efficient methodologies to provoke conceptual evolution and lead to an attitudinal change in the teacher's daily work with students is of great importance (ZULIANI, 2006, p. 14).

The aim of investigative work is for students to acquire knowledge that is compatible with scientific knowledge, to increase their observation of the facts of life, to start discovering problems in the things around them, and to dare to make guesses for their own investigations.

Being a guide in an investigation means giving hints that help students in their search for the knowledge they are looking for. It's up to the teacher to guide and indicate a path of enquiry, and not to keep giving students answers or suggesting ready-made knowledge.

Experimental teaching should be directly linked to the exploration of the new and the uncertainty of achieving success in what is being researched (experimented on), and should not be confused with a mere proof of theory, so as not to divide the classroom and knowledge. Moreover, only in practical classes do students face unexpected results. The failure of an experiment should also be addressed, clarifying what happened can be vital to increase the validation of the experimental lesson, making students look for solutions and explanations for the ineffectiveness, another very important point is the search for practices linked to the life and curiosity of students, whose understanding can serve as an application in their daily lives, because simply doing does not mean building knowledge and learning science (POSSOBOM et al, 2003).Carvalho (2004) states:

> Experimentation is essential for good science teaching, partly because the use of practical activities allows for greater interaction between teacher and students, often providing the opportunity for joint planning and the use of teaching strategies that can lead to a better understanding of science processes.

The same author argues that experimental activities should not be disconnected from theories, from group discussions; what was seen in class and what was obtained in the laboratory need to be constructed as something that complements each other. It often seems difficult, but it is fundamental, because experimental activities carried out without integrating theory do not allow for an effective understanding of what is being investigated.

According to Campos and Nigro (1999), investigations and experiments have some specific didactic purposes, such as: developing student autonomy; providing meaningful learning not only through conceptual change, but also through methodology and attitudes; enabling a view of science as an interpretation of the world, and not as a set of ready-made and defined answers; and broadly developing skills and capacities related to learning.

Francisco Júnior, Ferreira and Hartwig (2008) point out that the problematising experimental activity should give students the chance to carry out, record, discuss with colleagues, reflect, raise hypotheses and explanations, and discuss all the stages of the experiment with the teacher. This activity must be systematised and rigorous from the outset, awakening reflective, critical thinking in students and making them the subjects of their own learning.

Planning investigative experimental lessons is not always easy, but it must be done taking into account the student's interest, giving them the opportunity to participate in planning the activity and drawing up conclusions. This interaction allows laboratory activities to achieve the expected results, promoting discussion of the proposals and data collected. In this way, students become cognitively involved and exchange ideas between themselves and the teacher, which should lead to the construction of meaningful knowledge. Silva (2011) argues that practical work becomes more relevant when teachers are clear about its purpose and when they plan activities so that these purposes are achieved.

Silva (2011) also proposes a set of actions that could help teachers in their teaching planning; these are: analysing and selecting the content, pointing out possible difficulties in learning this content, considering the students' previous ideas and cognitive demands, selecting the teaching and learning objects and establishing the implications of the sequence and strategies chosen for teaching the selected content.

In order to carry out what was proposed in the planning, a teaching strategy is needed for the teacher's performance in the classroom. The teacher must cause an initial problematisation, where in the face of the new theme he presents a problem to the students in order to get to know their ideas and provoke a discussion of the subject in the face of their reality, so the student will have to look for other information to answer the initial question, systematising under the guidance of the teacher so that the organisation of knowledge occurs, always being attentive to what should be explored, contributing to the interpretation of the context so that the student realises that knowledge is accessible to all.

> Planning activities of this nature, given our educational reality, represents a significant change in the attitude of the teacher as well as the student and, consequently, in the dynamics of interactions in the classroom (SILVA, 2011, p 45).

Thus, we can consider that the teacher must create an environment in which students can think about their own ideas, modifying them with the contributions of their classmates and the teacher himself.

In view of the expansion of experimental activities of an investigative nature, it is considered that the teacher can acquire training to know how to propose situations and establish questions that can mobilise the students' previous knowledge, generate interactions between them and the teacher who guides the explanation of the data obtained in the experimental activities and the construction of scientific knowledge.

4.3 Between theory and practice - Teaching challenges - NEEDS

Today, science teaching is based on a constant relationship between theory and practice, between scientific knowledge and common sense. Always bridging the gap between what students learn in the classroom and what they experience in their daily lives. Delizoicov et al. (2007) point out that science is no longer a knowledge that takes place exclusively in schools, nor is its domain restricted to a specific layer of society, but rather affects humanity as a whole and each individual in particular.

Theory and practice should be seen as a single process, where there is no immutable truth, but rather theory serves to organise events and experiments, adapting theory to reality, as Giani (2010) explains when he states that there should be no distinction between the classroom and the laboratory, since, faced with a problem, the student would go beyond simple observations, as the hypotheses created should be discussed in order to assess feasibility and, if necessary, verify the different proposed solutions.

This learning is often hampered by the lack of integration between the subjects in the school curriculum, the physical conditions, the nature of the content, the poor qualifications of the teachers, the lack of equipment, the large number of students as well as their characteristics (BUENO and KOVALICZN 2009; SILVA and ARAÚJO 2011).

These problems in science teaching are made even worse by the fact that we find ourselves in a totally technological world, where these advances are growing on a large scale, and students are advancing along with them, but when it comes to teachers, we can say that many are still in the 18th century, with chairs lined up one behind the other, where students only play the role of listeners, silent, almost never intervening in class, but we know that today they participate, they give their opinions, so it is up to the teacher to reinvent himself in order to keep up with this advance and provide interactivity with the students. As Delizoicov et al (2007, p 33) state, not only has the student contingent increased, but also because the socialisation, forms of expression, beliefs, values, expectations and socio-family context of the students are different.

The school is part of this changing world, but what is still seen in the classroom are teachers who follow the textbooks to the letter. As Delizoicov et al (2007, pl27) confirm, the majority of natural science teachers still follow textbooks, insisting on memorising isolated information, believing in the importance of traditionally explored content and in exposition as the main form of teaching. Still on the subject, the authors point out that the training of science teachers also seems to have failed to take account of the change in the profile of school pupils, especially in primary education.

> In order to carry out experimental activities, science teachers need prior technical knowledge and to be able to handle different types of glassware, equipment, reagents, toxic substances and contaminants. It is known that many undergraduates have not always been adequately prepared to carry out laboratory activities, as they have often only participated in practical lessons in a passive way (BUENO and KOVALICZN 2009, p. 6).

Giani (2010) points out that the centrepiece of the lack of experimentation is teacher training and not just the lack of infrastructure.

> We believe that a well-structured laboratory will be of no use if teachers continue to take a simplistic view of experimentation, considering the exclusive functions of experimental work to be proving laws and theories, motivating students and developing technical or laboratory skills (GIANI 2010, p 20).

Another reason for not holding experimental classes is the lack of equipment needed to carry out experimental work, which for many teachers is considered essential and indispensable, as the school should have at least one space with some equipment and devices, but in its absence, this activity should not be left aside, the teacher should look for alternative means and materials to carry out the classes, as stated by Sereia and Piranha (2010, p 2):

> Sophisticated equipment and apparatus are not necessary for practical laboratory

> lessons. In the absence of such equipment, it is possible, according to the reality of each school, for the teacher to adapt their practical lessons using existing materials and also to use low-cost, easily accessible materials.

Experimental activities should not be limited to laboratory space with conventional materials. Some experiments can be perfectly carried out with alternative materials and spaces; this can even help to develop other skills, such as the ability to select and make use of unconventional materials (GIOPPO et al 1998).And finally, another point that causes great distress among teachers is the large number of students per class, which often makes it impossible to carry out the practices, because, despite being an important tool, as Bueno and Kovaliczn (2009) say, carrying out experiments in science is an excellent tool for students to make the content more concrete and to establish a relationship between theory and practice. Making it impossible with this large number of students, motivating them, suggesting a problem question and offering conditions for them to raise and test their hypotheses.

CHAPTER 5

METHODOLOGY

5.1 Characterisation of the School and Participants.

The Papa Paulo VI Municipal Primary School is located in the town of Nova Floresta in the state of Paraíba and caters for primary school students. It is a public school and has a laboratory that can be used by all subjects. The laboratory has some materials, such as: a uniocular microscope, an anatomical model of the human skeleton, an anatomical model of the male pelvis, an assemblable DNA structure, an anatomical model of the mouth and teeth, a eukaryotic cell with its organelles well delimited, a planetarium, as well as some games and maps.

The proposed activities were carried out at the start of the 2013 school year with two Year 7 classes° made up of 70 students aged between 11 and 17.

The experimental investigative activities were divided into three themes/areas of science content: Microbiology, Evolution and Plant Anatomy and Morphology.

Considering the conditions at the school, a script was drawn up for each experimental activity in order to help them carry out the activities. The script contained: the main objective of the lesson, a short text for the student on the subject to be discussed, the material to be used in the activity and the procedure, i.e. the process that would be required to carry out the experiment.

5.2 Research instruments

In order to collect the data for this research, the students' prior knowledge on the topics of the investigative lessons was assessed. In order for the data to be collected in the best possible way, the research material was collected using various instruments, including: a previous questionnaire on the topic of the investigative activity, an experience notebook and finally the re-use of the previous questionnaire, now to assess what had been learnt in the experimental investigative activity.

For the sake of organisation, the research instruments, as well as the specific procedures for applying each one, will be described in the same order as they were presented to the students.

5.2.1 Questionnaire on prior knowledge

The first survey instrument used was applied individually to each student. The instrument was made up of a series of discursive and multiple-choice questions, where each question consisted of an assessment of the student's understanding of the topic.

All the students in the same class answered this questionnaire, always at the first meeting of each subject/area. As it was the first data collection instrument, the students were told about the general aim of the research before it was administered.

The application of the questionnaire was important to determine whether or not the students had mastered the content that the experimental investigative activity would deal with.

5.2.2 Experience booklet

The second data collection tool used was an experiment notebook, which each student received at the first meeting, where at the end of each experimental activity they would report what they had done in class and leave their hypothesis, as well as answering some of the questions posed during the activity.

When describing the activity, care was taken not to suggest or influence any form of student reasoning. This stage is part of the process of hypothesising, allowing students to explain their ideas about the experiment.

5.2.3 Questionnaire to reassess the student's knowledge of the subject/area

In this phase of data collection, the students again answered the questionnaire proposed at the first meeting, only this time the questionnaire was designed to assess their understanding of the content after the experimental lesson. This is a way of estimating whether there has been any growth in relation to the content, taking into account the theory-practice approach.

In this phase of data collection, it was possible to see that the students explained and dealt with the questions more spontaneously and confidently, and it was clear that they were getting closer to the content.

5.3 Procedures

The experimental activities and the collection of the research material took place from 18 February to 27 March 2013. The experimental classes were held during normal school hours, as the school was in the process of adapting to the Comprehensive School System. The 70 students taking part in the research were divided into two classes. The meetings were always held in two lessons lasting 60 minutes each. The order of the topics was the same for both classes. The first topic was Microbiology, followed by Natural Selection and Plant Anatomy and Morphology.

Most of the students (except those who missed a particular lesson) answered the questionnaires and carried out the experimental activities on the three themes. The investigative experimental activities were carried out in groups, with the number of students varying depending on their attendance. The choice of groups was made in an attempt to bring the students closer together, especially at times of questioning and hypotheses that would arise during the experimental activity, as well as to better organise the large number of students, given the space that is available to carry out such activities.

5.4 Descriptions of the experimental investigative lessons

Topic of the experiment: MICRORGANISMS

I° Meeting - CULTIVATING MICRORGANISMS

This activity was taken from Nova Escola magazine and aimed to show the existence of microbes and how they contaminate the culture medium. Firstly, the students were explained what the project was about, and then they were given a questionnaire containing 14 questions, 05 of which were discursive and 09 multiple choice, to analyse their prior knowledge of the subject of the lesson. The questionnaire can be found in Appendix A of this paper. After everyone had answered the questionnaire, a brief explanation of the topic began, where many had seen or heard about microorganisms for the first time.

Materials:

- Colourless gelatine,
- Meat broth,
- Water,
- Swab,
- Plastic film,
- Tags,
- Pens,
- Test tube,
- Petri dish.

Procedure:

- After explaining the procedure for the experiment, the students were shown how to develop a homemade culture medium using accessible materials such as gelatine, meat broth and water. The culture medium (Mueller Hinton Agar) used in the classroom was made in advance at the Cuité Campus - UFCG, using materials provided by the school, as the school did not have the materials or the fridge to preserve the medium.
- Divided into groups of 5 students, each group was given a Petri dish containing the culture medium, a test tube and a cotton bud, where they had to look for areas where they thought microorganisms might be.
- After collecting the microorganisms, the students returned to the classroom and were instructed to use the "dirty" cotton swab to contaminate the culture medium, which was wrapped in plastic film and labelled with the name of the group and the place where the microorganisms were collected.
- After the experiment, each student was given a notebook so that they could report on the lesson and raise hypotheses about the experiment so that they could later compare it with the real result.

2º Meeting - TESTING CLEANING PRODUCTS

Before we started the second meeting, each group had 5 minutes to observe their culture medium and confirm or not their hypotheses. After the observation, the students wrote down what they had observed, taking into account their previous conceptions.

This activity was taken from Nova Escola magazine and aimed to prove the effectiveness of disinfectants and other products that promise to kill microorganisms. Initially, each group received their own culture medium so that we could use it to carry out the lesson's procedures.

Materials:

- Bacteria from the first experiment,
- Petri dish with culture medium,
- Paper filter,
- Tweezers,
- Test tube,
- Cleaning products (bleach, disinfectant, detergent and multi-purpose)
- Greenhouse.

Procedure:

- Each group scraped some of the bacteria from the already contaminated plates and diluted them in a few drops of water in the test tube, after which they spread the mixture of water and microorganisms in the Petri dish with the culture medium.
- Using tweezers, a representative from each group dipped the paper filter into the cleaning materials, and each group had the autonomy to choose which cleaning material to use.
- The paper filter with the cleaning product was then placed in the middle of the plate contaminated with microorganisms and kept in the oven.
- At the end of the experiment, the students were encouraged to write down their suggestions for the next lesson and whether or not the cleaning products were effective.

3º Meeting - CONCLUSIONS

At the last meeting on the subject of microorganisms, each student was able to observe with their group the results of the experiment carried out in the previous lesson. They observed the culture media present in the greenhouses and the results obtained by the action of each cleaning product, and each student received their notebook in which they confirmed or not the effectiveness of the cleaning product. After the report, the students were given the questionnaire again, this time to answer the questions after their experience of the subject, so that we could see if the students had understood the content. Based on the number of questionnaires obtained at this meeting, we realised that the number of students had decreased compared to the first meeting, as many students had left the school due to the implementation of the Comprehensive School System.

Experiment theme: NATURAL SELECTION

Iº Meeting - THE TENTILIONS OF GALAPAGUS

This activity, based on the article "The finches of the Galapagos: what Darwin didn't see but the Grants did", was designed to simulate what happens in nature in relation to the availability of food. Firstly, the students were given a script of the lesson, then some oral questions were asked to check the students' knowledge of the subject. As the students hardly spoke, we preferred to give an introduction to the content before the experiment, then the students were divided into groups of five to carry out the experiment.

Material

- Plastic pot containing various seeds;
- Transparent plastic tray;
- Pointless scissors;
- Nail pliers;
- Eyebrow tweezers;
- Clothes peg

Procedure

- Each group was given a tray containing different types of seeds, 5 tools (2 nails, 1 pair of pliers, 1 pair of scissors and 1 pair of eyebrow tweezers) for each group member to choose their tool and 5 disposable cups for each student to store the seeds they had collected.
- Each group collected the seeds for 30 seconds using the instruments and not their hands. Each student kept their seeds separate, and after five rounds of 30-second collection, they checked which seeds had been collected and which were left over;
- If a student didn't manage to get any seeds, they wouldn't be able to take part in the next round;
- At the end, each student was given a notebook with a few questions about the experiment so that we could check whether or not they had understood the relationship between the instruments and seeds and the adaptation of the Galapagos finches. We also carried out a survey of the sizes and quantities of seeds collected by each type of "beak" in each round. The questionnaire for this experiment can be found in Appendix B.

Topic of the experiment: PLANT MORPHOLOGY AND ANATOMY

Iº Meeting - SEED OR NOT

This activity was based on the article "A seed, a plant?" and its aim was to determine what a seed is. Firstly, the students were told what the topic of the lesson would be, and then they were given a questionnaire containing 10 questions, 5 of which were discursive and 5 multiple choice, to analyse their previous knowledge of the subject. The questionnaire can be found in Appendix C. After everyone had answered the questionnaire, a short debate began on what is a seed? What is inside it that makes it become a plant? What conditions are needed for a seed to germinate?

To encourage the students in their thinking, they answered the following question: how do you check if they are seeds or not? Faced with this question, many students suggested that the seeds should be sown in order to check.

In response to this suggestion, a collective oral discussion began with the aim of getting the students to formulate their predictions, and various questions were used to help the students with their reasoning.

- What happens if you sow all this?
- If they grow, does that mean they are seeds?

- The students answered these questions in their notebooks by diagramming the experiment and anticipating the result, explaining the reasons. The students organised their planting in groups, planting at a certain distance from each other and identifying each peti bottle with the name of the group.

2º Meeting - OBSERVATION OF PLANTING

The aim of this lesson was to observe and draw conclusions from the plantings made in the previous lesson. Initially, the students were distributed in the groups from the previous lesson, where each group could observe their plantings and draw conclusions based on the hypotheses raised in the previous lesson. As the time between the first meeting and the second was short, almost nothing grew, but this was important for the students to come up with some suggestions to explain what they had seen.

After these observations, the students were asked to reflect on the experience, based on a few questions:

- What has grown?
- What hasn't grown?
- How can we explain this relationship between a seed and a plant?
- Why haven't some seeds germinated yet?
- n the next meeting, will anything else have grown?

Based on these questions, the students were encouraged to write down in their experiment notebooks the reasons why some seeds didn't germinate.

3º Meeting - SEED ANATOMY

At the start of the lesson, each group was taken to observe their planting, now with some sprouts already growing. For the seeds that hadn't germinated, some were chosen for observation of the inside of the seed with the naked eye and then with the help of a magnifying glass, each student drew what they could see with the naked eye and then with the help of the magnifying glass.

Next, an expository lesson was given using slides to explain the anatomy of the seed, as well as its parts and functions. After the explanation, students were asked some questions to reflect on the topic.

- What does the seed need to germinate?
  - What makes certain seeds grow faster than others?
  - The need for water and light for germination will also be discussed

The students were then asked to plant seeds with water and without water, with light and without light, in order to check some of the conditions necessary for germination.

Material

- Disposable cups
- Water
- Cotton

Procedure:

- The same groups as in the previous experiment were asked to sow bean and maize seeds in disposable cups with cotton wool moistened with water or not. Each group chose a place in the laboratory to plant their seeds, bearing in mind that some had to plant their seeds away from light and others in contact with light.

4º Meeting - CONCLUSIONS

At this meeting, the groups observed the plantings made in the previous lesson, taking into account the germination conditions explored by each group. After the observation, each student reported the results of the experiment in their notebook, making an interaction between the presence or absence of water and light as necessary factors for germination. At the end of the meeting, a questionnaire was administered so that this time, after the experiments, we could see if the students had understood the content

CHAPTER 6

RESULTS

6.1 EVALUATION OF LEARNING ABOUT MICROORGANISMS THROUGH EXPERIMENTAL LESSONS

As described in the methodology, in the first meeting on the subject of microorganisms, the students were encouraged to answer a questionnaire to find out their previous knowledge of the subject. The answers were grouped into right, wrong and no answers to check their initial ideas (Figure 1).

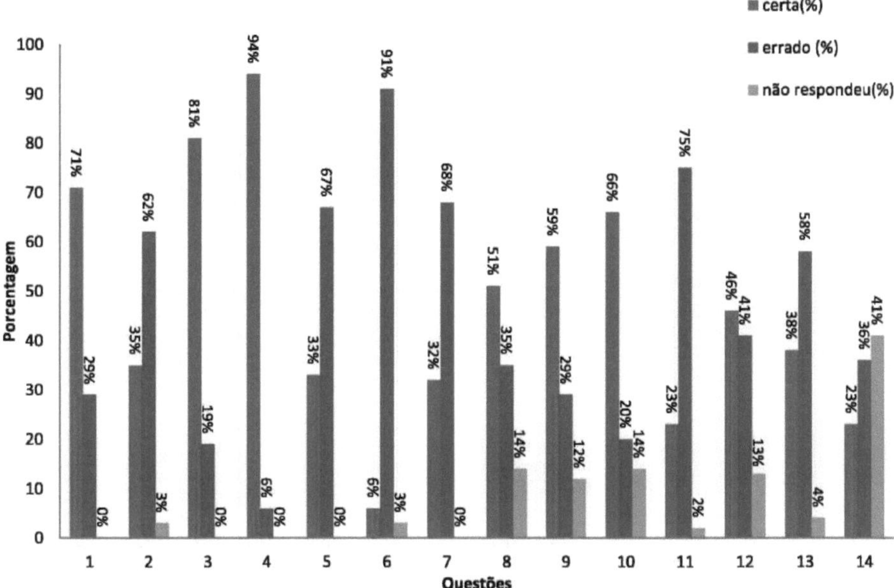

Figure 1: Results of the questionnaire on microorganisms administered to 69 pupils in year 7° of the Papa Paulo VI Municipal Primary School.

It can be seen in the graph that questions 05, 06, 07, lie 13 had a low hit rate. These questions dealt respectively with: the group of microorganisms, the use of microorganisms in cleaning environments, the curiosities of these beings, their functions and the factors necessary for their proliferation.

Questions 06 and 11 actually involve technical knowledge that is generally difficult for students to assimilate, but questions 05, 07 and 13 are relatively easy subjects that they often see in everyday life.

It can also be seen that a large number (41 per cent) of students did not answer question 14, even though it was a discursive question about the action of cleaning materials, thus showing that students feel insecure about answering simple questions related to their daily lives.

The experimental lesson on the cultivation of microorganisms was very interactive, all the students were willing to participate, divided into groups as described in the methodology, the students with the help of the cotton swab sought to collect microorganisms, back in the laboratory each representative of the group contaminated the culture medium with the supposed microorganisms, after the experiment each student received his experience notebook where he left his hypotheses for the following questions:

- How will the culture medium be in the next class? Will it grow anything?

To these questions, the students reported the following hypotheses:

"I think there's going to be a cake of bacteria in the lab next week. "

"I think it's going to have a lot of germs, bacteria and much more. "

"The collection was done in the bathroom and I think there will be lots of bugs in the glass. "

Figure 2. A student's conception of the I° meeting

In the second meeting, before tackling the cleaning materials experiment, the students had five minutes to observe and report on whether or not their hypotheses raised in the previous lesson had been confirmed.

The students reported this observation:

"It was the way I thought it would be, it was big and had lots of microorganisms and it was big and colourful, there was yellow, white and orange, I liked the white one best."

"I thought it was small but it's big, I also thought the bacteria were going to turn black but they've turned yellow. "

"In the bathroom the mould on the wall has turned black and ugly."

"I thought it was going to be pink like a triangle, but no, it turned yellow in the shape of a very large yellow circle. "

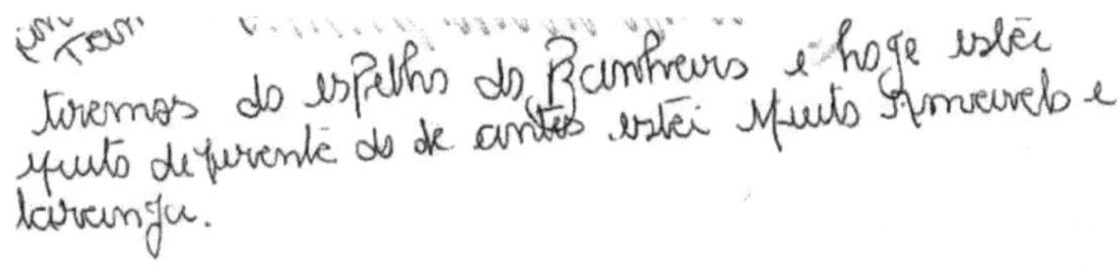

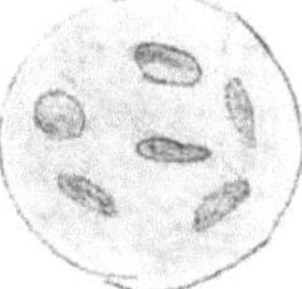

Figure 3: Confirmation of the hypothesis raised at the I° meeting.

In this second meeting, with the students split into groups, tests were carried out to observe the effects of different cleaning products on microorganisms, according to the procedure explained in the methodology:

> *"I think that the products we put on the bacteria, I think that when we go to look at the product it will have had an effect and killed the bacteria. "*
>
> *"We did an experiment with bleach, we wet the filter paper with the product and put it in the culture medium, I think the bleach will kill all the bacteria. "*
>
> *I don't think the detergent will kill much because it's used for washing dishes"*
>
> *"Well we put disinfectant in the culture medium full of bacteria to see if it kills the microorganisms as they say, I think it will kill them. "*

In the third meeting of this theme, the culture media were observed under the effect of different cleaning products. Each group had time to observe their plate and the plates of the other groups, noting which ones obtained the best results:

> *"Killed all the micro-organisms"*
>
> *"Detergent killed the bacteria and any chemical does. "*
>
> *"What killed our bacteria was the multi-purpose cleaner. "*

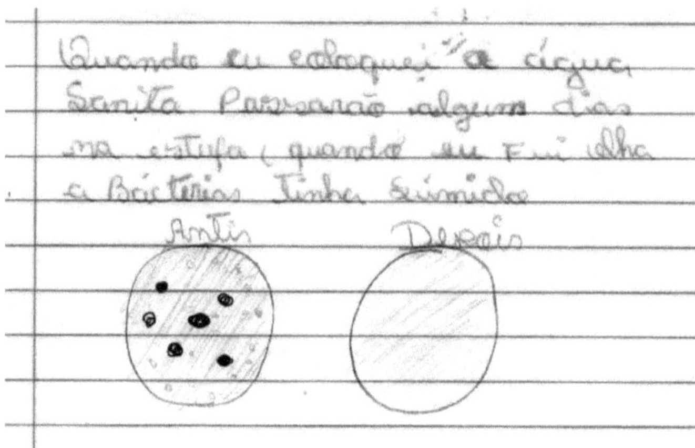

Figura 4. Observation made by a student in the third meeting.

Also at this meeting, the students returned to answer the questionnaire from the first meeting, only this time to assess their understanding of the content after the experimental lesson (Figure 5).

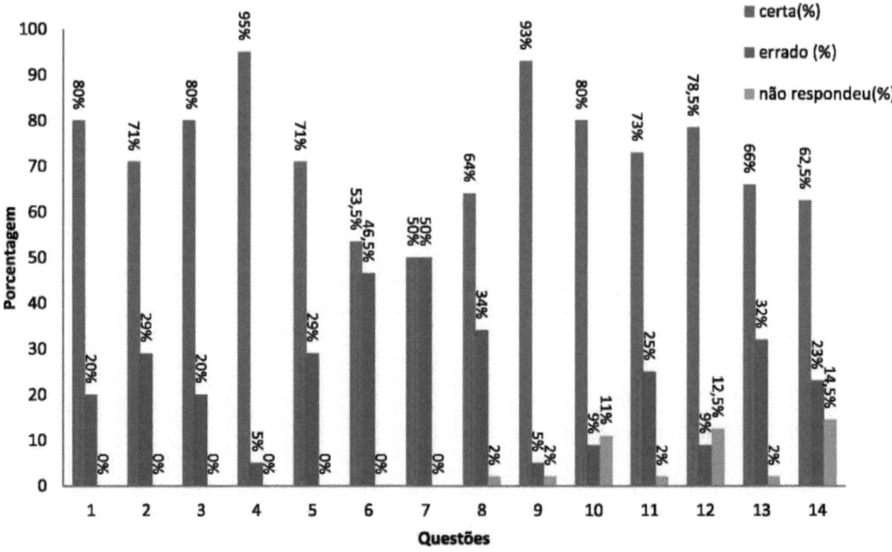

Figura 5. Questionnaire after experiments on microorganisms, given to 56 students in the 7th grade° of the Papa Paulo VI Municipal Primary School.

It can be seen that after the experiments and a brief explanation of the topic, the students were more confident in answering the questions related to their daily lives. It can now be seen that few of them stopped answering the discursive questions, making it clear that they understood and were confident in talking about the topic, as well as the higher percentage of correct answers to the questions after the experiments (Figure 6).

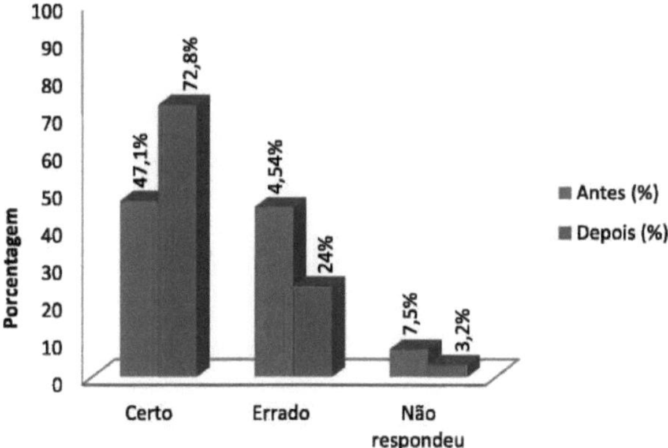

Figura 6. Comparison between the results of the questionnaire before and after the experiments.

6.2 EVALUATION OF LEARNING ON THE SUBJECT OF NATURAL SELECTION THROUGH THE EXPERIMENTAL LESSONS.

As described in the methodology, this experiment sought to relate the availability of food and the different types of beaks found on finches. As the students didn't express themselves when asked about the topic, they were given a brief introduction to Darwin's accounts of his trips to the Galapagos, then divided into groups, After the experiment, the students answered a questionnaire containing five discursive questions to check their understanding, relating the topic to the experiment. The students' answers were grouped into 3 categories: understood, partially understood and did not understand (Figure 7).

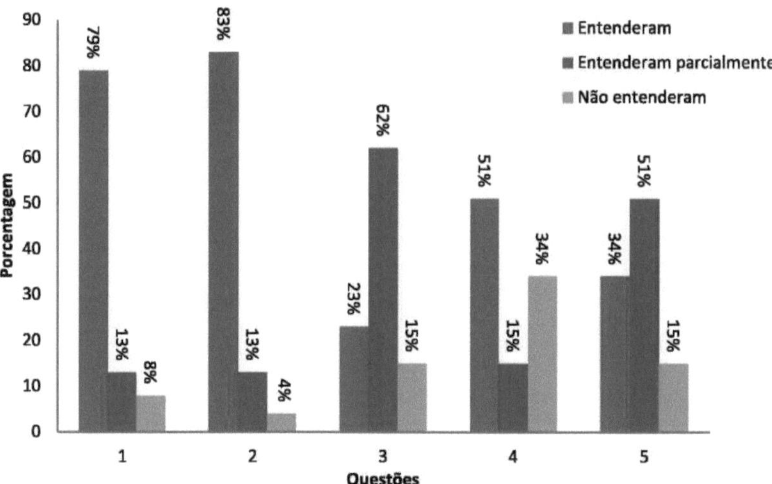

Figura 7. Verification of understanding by question of the experiment on Natural Selection applied to 53 students in Year 7° of the Papa Paulo VI Municipal Primary School.

The students' answers were analysed into three categories of understanding of the experiment: understood, partially understood and did not understand. For a better understanding of these criteria, a summary table will be described below, made up of the questions in the questionnaire and the answers considered correct and partially correct according to the experiment's objective. It is not necessary to describe the answers that the students did not understand, as these answers are completely different from the expected answers.

Table 1. Questions and answers according to the aim of the experiment.

Question	Correct answer	Partially correct answer
1. What does each of the tools used to collect the seeds represent?	"The beaks of birds"	"The tools to collect the seeds"
2. What does the tray with the seeds represent?	"Bird food"	"A bird eating"
3. What does each round of seed collection represent?	"The seasons"	"0 collection time"
4. What does the elimination of individuals who fail to collect seeds in a round represent?	"The death of birds"	"Birds that don't have the agility to eat"
5. Draw up a table to record the number and variety of seeds your instrument has managed to collect.	Answer made with table	Answer made without a table, just describing the seeds and their quantities.

Taking into account the categories mentioned above, it can be seen that the questions that had the highest level of understanding were questions 1 and 2, while questions 3 and 5 were partially understood, and only question 4 showed a significant number of students who did not understand its application, even though it was a simple question that could be easily understood through its application (Figure 8).

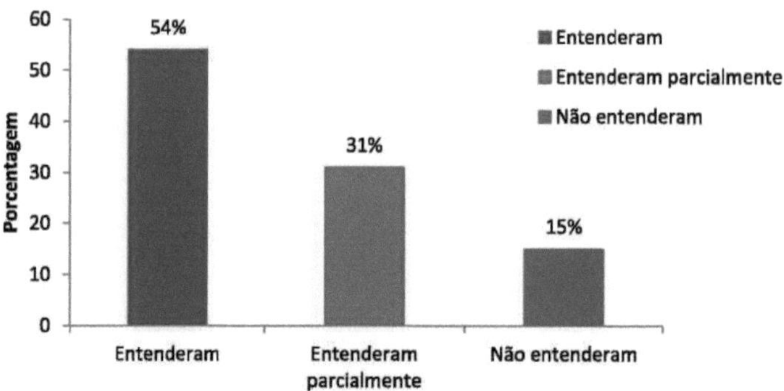

Figura 8. he overall result of the experiment on Natural Selection applied to 53 students in the 7th grade° of the Papa Paulo VI Municipal Elementary School.

6.3 EVALUATION OF LEARNING ABOUT SEEDS THROUGH EXPERIMENTAL LESSONS

In this first meeting on the subject, the students were initially encouraged to answer a questionnaire to find out their prior knowledge. The students' answers were grouped into right, wrong and no answers, in order to ascertain their initial ideas (Figure 9).

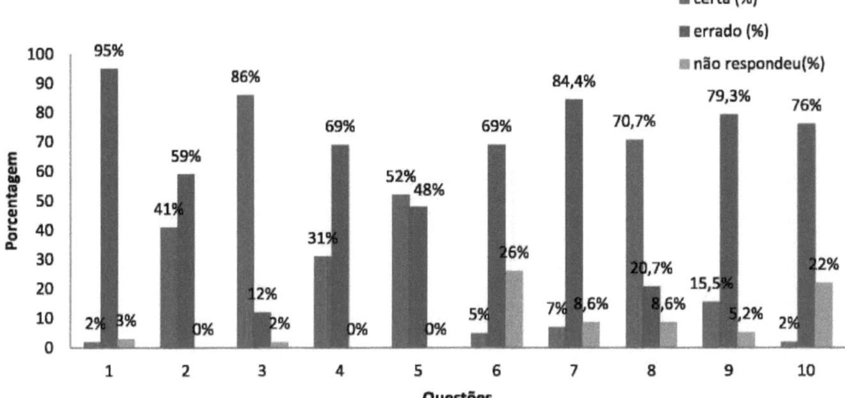

Figura 9. Preliminary questionnaire on seeds given to 58 pupils in the 7th grade° of the Papa Paulo VI Municipal Primary School.

The graph shows that questions 1, 4, 6, 7, 9 and 10 had a very high error rate. These questions dealt respectively with the definition of seed, germination, parts of the seed, functions of the seed, functions of the root and the importance of the seed in adapting to the terrestrial environment.

Some questions, such as 6 and 10, are discursive and require a little more knowledge on the subject, but the other questions are simple and part of their day-to-day lives. In relation to question 1, many students said that a seed was a plant or something that, when planted, grew and bore fruit, This way I could see that many students were generalising their knowledge of plants with the topic of seeds, while question 4 about germination, although many had already observed this process, they were unable to relate the word germination to the initial process of plant growth, something that is commonly done by several students in the class who live in the rural area of the municipality of Nova Floresta.

Questions 02 and 08 had the highest number of correct answers, as they dealt with seed dispersal and the elements needed for plant growth.

After solving the questionnaire, the students were asked to observe some grains and classify them as seeds or not. Divided into groups, the students were asked to classify their grain samples as seeds or non-seeds.

Then the following question was asked with the aim of proving this classification: How do you check if they are seeds or not? It didn't take long and some of the students suggested planting the grains considered to be seeds. Each group was given a peti bottle that had been cut open and a certain amount of fertiliser to plant them in. After the experiment, each student received an experiment notebook in which they wrote down their hypotheses for the following questions:

- What will our planting be like in a few days' time?

To these questions, the students reported the following hypotheses:

"I don't think many seeds will grow. "

"I think there's going to be a very big plant, but very small. "

"I think the eye of the bean and the corn will come out. "

"I think some seeds will be big, but there are some that won't grow, they'll be beautiful. "

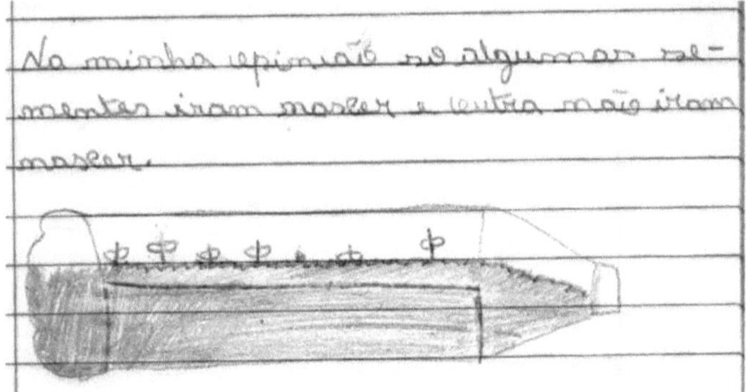

Figure 10. Hypothesis raised by a student at the I° meeting on seeds.

The second meeting was devoted to observing the planting. Each student was encouraged to observe their group's planting and report in their experience notebook what they had observed, raising arguments for their suggestions.

The students reported this observation:

"It didn't grow due to lack of water and electricity. "

"It didn't grow because we put a lot of water on it. "

"I don't think it grew because it was only a few days. "

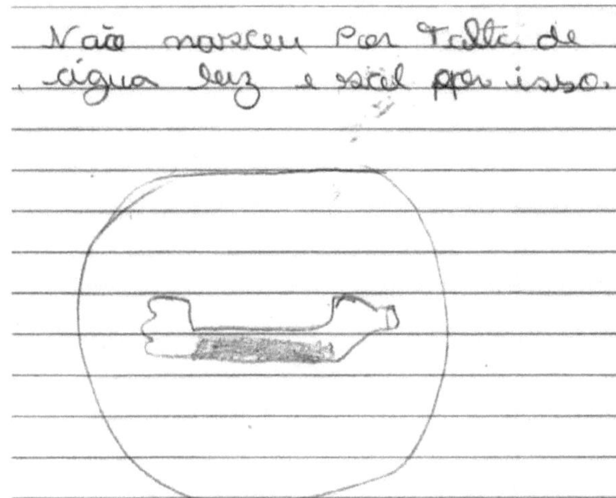

Figura 11. Suggestion raised by a student during the planting observation lesson.

In the third meeting on seeds, the students were once again encouraged to observe their planting, now with some sprouts already growing. For the seeds that hadn't germinated, some were chosen to observe the inside of the seed with the naked eye and then with the help of a magnifying glass. After observing the inside of the seeds, a lecture was given on the anatomy and morphology of the seed, explaining its parts and functions. At the end of the presentation, the students were asked about the conditions necessary for germination, whether or not they needed water and light.

Next, the students were asked to plant seeds in cotton cups again. Some cups were given water and others not, and others were exposed to light while others remained in the dark, so that we could check some of the conditions necessary for germination.

For this experiment, the students reported:

"Five little plants came up, how nice! I think the others didn't grow because we didn't water them properly and today we put beans to germinate in the cotton but without water, I don't think they'll grow at all."

"From the previous experiment, the seeds grew, not many, but today's I think that in my glass next week it will transform, the embryo is protected by the shell, so next week I think it will transform. "

At the last meeting on seeds, the students were able to observe their plantings in cotton wool, and the cups that received water and light germinated, thus confirming some of the conditions necessary for germination.

The students reported this observation:

"The seed germinated because we put water on it and the only reason it hasn't grown more is because it's not in the ground. ".

" It didn't grow because it was in cotton wool and we put it in the dark and without light the seeds can't germinate. "

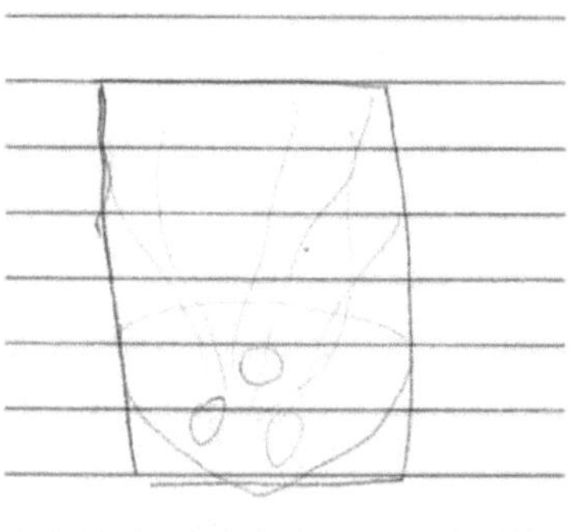

Figura 12. Illustration by a student showing the germination of seeds that have received water and light.

Also at this meeting, the students returned to answer the questionnaire proposed at the first meeting, only this time to assess their understanding of the content after the experimental lesson (Figure 13).

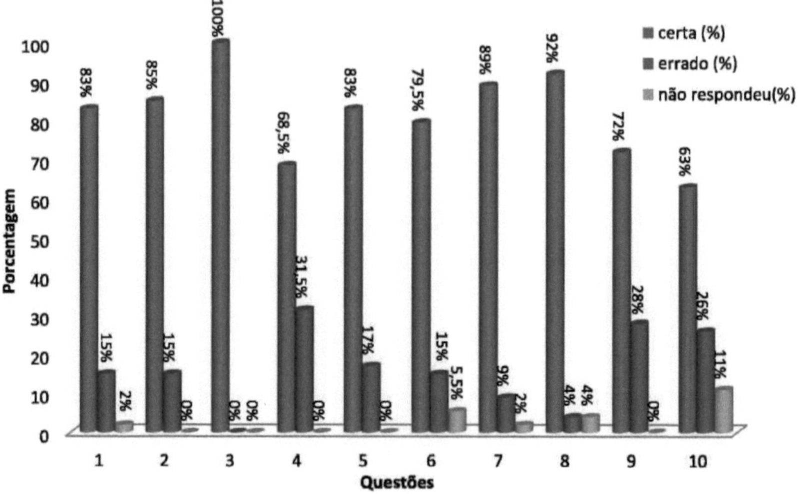

Figura 13. Results of the questionnaire after experiences with seeds, given to 54 students in year 7° of the Papa Paulo VI Municipal Primary School.

It can be seen that after the experiments and the lecture, using resources such as images and drawings on the subject, the students were more confident in answering the questions related to their daily lives, it can be seen that almost no discursive question was left unanswered, thus verifying the understanding of a content so present in the lives of many students, in addition to the significant drop in the number of errors related to simple questions. (Figure 14.)

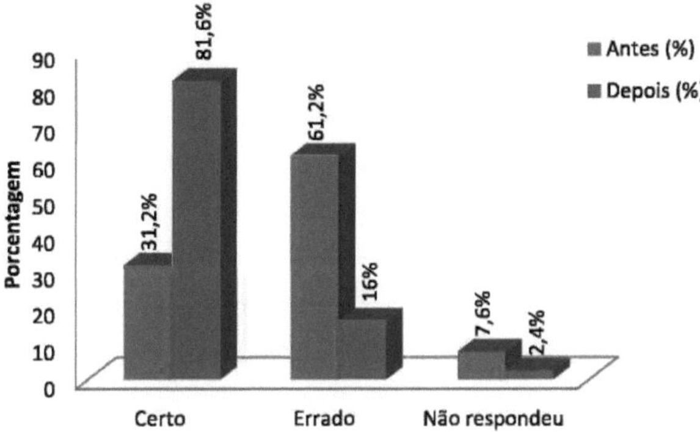

Figura 14. Comparison between the results of the questionnaire before and after the experiments.

CHAPTER 7

DISCUSSION

According to Melo (2010, p. 41), laboratory activities should adopt procedures based on a pedagogy aimed at not providing a methodology that stifles the experimental process, making it purely mechanical, but rather a practice that starts from the student's experience, so that what is being proposed makes sense in the student's universe.

In an effort to understand the students' prior knowledge, a questionnaire was administered at the beginning of each experiment to observe their understanding of the topic. These questionnaires showed that the students initially felt insecure about answering the discursive questions, often leaving the questions blank and when they did answer, their responses were inadequate. This made it clear that they had little or no knowledge of certain subjects. These observations are in line with the work done by Cabrera (2007), who describes that some of the content analysed in the prior knowledge questionnaire was not adequately justified, and some students were unable to express their opinions appropriately in relation to the topic.

Also according to Cabrera (2007), the results of the questionnaires relating to prior knowledge were important for planning and designing playful instructional strategies to facilitate interaction between new and pre-existing information, with the aim of promoting meaningful learning.

It was possible to consider the prior knowledge questionnaire as an important resource for determining the steps to be taken, as well as the shortcomings to be overcome, through the investigative experiments.

Investigative experimental activities were an important tool used to promote the construction of meaningful knowledge by students, as these lessons arouse students' interest, motivating them to investigate and seek solutions to problems around them (SUART et al 2009; SEREIA and PIRANHA, 2010).

Corroborating Sereia and Piranha (2010), we realised that the students were more participative and interested in investigative practices.

In building meaningful learning, Cabrera (2007) also used participation and construction in a relaxed environment as a methodological strategy, working collaboratively, negotiating, interacting and reflecting to promote cognitive development.

In order to see if there had been any growth in understanding of the themes after the experiments, the same questionnaire was administered as at the first meeting, with the aim of tracing the students' approach to the content. This time, few students left the questions blank, and it was also noticeable that the students answered the questions spontaneously and confidently, thus increasing the number of correct answers.

°This corroborates the results of Filho et al. (2009) in a study carried out with secondary school students in a public school in Marília/SP, using a questionnaire to compare the answers before and after the activities. It was possible to see a significant increase in the percentage of correct answers after the activities, where the author reports that it was possible to gradually introduce new concepts to the students in a constructivist approach.

[a]The data obtained also corroborates the observations made by Prigol and Giannotti (2008) in their work with 6th graders at a state school in Cascavel, PR, where they applied a questionnaire beforehand and afterwards to assess the students' knowledge, and found that those who took part in the practical classes had the highest number of correct answers (from 39.6 per cent beforehand to 70 per cent after the practical lessons).

Borges (2001) states that investigative laboratories can be used by any level of education and have the potential to involve students more, leading to an improvement in student responses.

In view of the results obtained, the importance of experimentation in science teaching became clear, especially when it deals with subjects from everyday life, because this is the only way for students to feel that they are in a single world and not in two (school x everyday life). It was possible to realise that these activities improved the students' knowledge, enabling them to develop the ability to discriminate an appropriate concept from an inappropriate one for the same situation.

The students' involvement with these perspectives was only possible due to their interaction and motivation while carrying out the investigative activities in pursuit of meaningful learning. In this environment, the students not only generated their own hypotheses and arguments, but debated them with their classmates and teachers.

This is in line with Coll et al's (2006) assertion that the teacher must intervene to activate the students' previous ideas, allowing them to learn the objective of the task or experiment to be carried out, as well as encouraging the students to summarise, take notes, draw up work plans, discuss and establish moments for monitoring and self-evaluation of the process.

According to Melo (2010), the laboratory encourages students to know, understand and learn to apply theory in practice, mastering tools and techniques. It enables students to learn to observe scientifically, interpret and analyse instruments through objectivity, precision, confidence, perseverance, satisfaction and responsibility. This was confirmed in their work after applying a new methodology to practical classes, where the students interviewed said that laboratory classes made it easier to learn, improving their understanding of the subject.

It was therefore clear from the methods used that the students assimilated the content better when there was an interaction between theory and practice.

This corroborates Gomes (2005) who, based on the results of his research, states that it is only by understanding what students think and do during laboratory activities and the potential of an enquiry-based learning environment that it is possible to promote a greater epistemological understanding of science.

According to Tapia and Fita (2004), learning is the change that takes place in the student as they move from an initial and elementary state to a more structured and organised final state.

In this learning proposal, the students were able to make progress in relation to the topics investigated through the exchange of information, ideas, questioning and problem-solving. It is clear to see the importance of using these activities, which, when well planned, make it much easier to understand the production of knowledge in science teaching.

CHAPTER 8

FINAL CONSIDERATIONS

Investigative experimental lessons are an important strategy that can be used by teachers to diversify their daily practices, and when used in the right way, they contribute greatly to meaningful learning. In this form of teaching, unlike traditional lessons that are approached using only textbooks, it is possible to promote the development of students' autonomy and ability to make decisions, to evaluate and solve problems so that students can get closer to the concepts and theories of science teaching.

In inquiry-based science teaching, students interact, explore and experiment with the natural world, and are not restricted to purely mechanical manipulations or observations. They are inserted into investigative processes, getting involved in their own learning, constructing questions, developing hypotheses, analysing the results, drawing conclusions and reporting the results. In this approach, learning goes beyond the mere execution of activities and becomes an opportunity to develop new visions, meanings and knowledge.

In this sense, teachers have a key role to play in encouraging and motivating students, guiding them through their suggestions and doubts so that they don't get lost in their analyses, always focusing on the problems to be explored and solved.

In this way, as Cabrera (2007) states, the pleasure of studying through valuing thought motivates students to learn more meaningfully and the promotion of a relaxed atmosphere avoids laziness, disinterest and discouragement, as well as promoting the pleasure of studying and learning.

Thus, it can be seen that there was a significant increase in the teaching-learning process, in terms of the level of understanding of the content and the interaction of the students in class after the investigative experiments, given the need to organise the students' knowledge and ideas in order to solve the problem questions.

Through the inclusion of basic concepts, where the students themselves constructed the knowledge and drew their own conclusions, a growth in meaningful learning was realised, moving away from traditional teaching, where students would be mere recipients of knowledge, thus becoming the agents of learning themselves, making it clear how important it is to invest in carrying out experimental activities in schools. As Melo (2010) states, "what students only hear they soon forget, what they see they keep little of, but what they do and describe they hardly forget".

Given this statement and the results found in this research, it can be concluded that experimental lessons in science teaching are stimulating and fundamental for students to build their knowledge. In these classes, students are able to reason better, reflect and relate what they were researching or studying in the classroom to their everyday lives.

REFERENCES

ARAÚJO. M.P, RODRIGUES. E.C, DIAS M.P. A, FARIAS M.J.R. **As atividades experimentais como proposta na abordagem contextualizada dos conteúdos de biologia,** 2012. Available at: http://www.nutes.ufq.br/abrapec/viiienpec/resumos/rl386-l.pdf. Accessed on: 20 January 2013.

BEVILACQUA, G. D. AND COUTINHO-SILVA. R. (2007). Science teaching in the 5th grade[a] through experimentation. **CIÊNCIAS & COGNIÇÃO,** 2007; Vol. 10. Available at: http://www.cienciasecognicao.org/pdf/vl0/m317138.pdf. Accessed on: 18 October 2012.

BIZZO, N. **Ciências: fácil ou difícil.** Ed. Ática, São Paulo, SP, 1998.144p.

BORGES, A.T; BORGES, O.N; SILVA, M.V.D; GOMES, A.D.T. **The resolution of Practical Problems in the School Laboratory.** Atibaia, SP: 2001. Available at: http://www.nutes.ufq.br/abrapec/iiienpec/Atas%20em%20html/ol3.htm. Accessed on: 19 November 2012.

BRAZIL. **National Curriculum Parameters: Natural Sciences.** Brasília: MEC/SEF, 1998. Available at: http://portal.mec.gov.br/seb/arquivos/pdf/ciencias.pdf. Accessed on: 13 January 2013.

BUENO, R.S.M.; KOVALICZN, R.A. **Science teaching and the difficulties of experimental activities.** 2009. Available at: <http://www.diaadiaeducacao.pr.gov.br/portals/pde/arquivos/23-4.pdf.> Accessed on: 12 August 2011.

CABRERA, W. B. **A Ludicidade para o Ensino Médio na disciplina de Biologia: Contribuição ao processo de aprendizagem em conformidade com os pressupostos teóricos da aprendizagem significativa.** 2006. Dissertation (Master's Degree in Science Teaching and Maths Education). State University of Londrina, Londrina, 2007.

CAMPOS, M.C.C. E NIGRO R.G. **Didática de Ciências - O Ensino-aprendizagem como investigação.** São Paulo, FTD, 1999.

CARVALHO, U.R.L; PEREIRA, D.D; MACEDO, E; SILVA, K; CIBELE, M; FOLENA, M. A Importância das Aulas Práticas de Biologia no Ensino Médio. **JEPEX, UFRPE,** Recife, 2010. Available at: http://www.trabalhosfeitos.com/ensaios/a-Import%C3%A2ncia-Das-Aulas-Pr%C3%Alticas-De/414896.html

CASTOLDI, R.; POLINARSKI, C.A. The use of didactic-pedagogical resources to motivate learning. **I NATIONAL SYMPOSIUM ON THE TEACHING OF SCIENCE AND TECHNOLOGY,** 2009.

CENTRE FOR SCIENTIFIC AND CULTURAL EDUCATION (CDCC) - USP. Teaching science at school, from kindergarten to fourth grade. São Carlos, 2005.

COLL, C. **Constructivism in the Classroom.** São Paulo: Editora Ática, 2006.

DELIZOICOV, D.; ANGOTTI, J. A.; PERNAMBUCO, M. M. **Ensino de Ciências: Fundamentals and Methods.** São Paulo: Cortez, 2007. UNESP - Botucatu Campus

FARIAS, C. S.; BASAGLIA, A. M.; ZIMMERMANN, A. The importance of experimental activities in chemistry teaching. In: **Iº CPEQUI - Iº CONGRESSO PARANAENSE DE EDUCAÇÃO EM QUÍMICA.** 2009.

FRANCISCO JR, W.E.; FERREIRA, L.H.; HARTWIG, D.R. Experimentation Problematising: Theoretical and Practical Foundations for Application in Science Classrooms. **QUÍMICA NOVA NA ESCOLA,** n°30, 2008.

GENTILE,P.A. Como ensinar microbiologia com ou sem laboratório. ***Nova Escola,*** São Paulo, n. 183, June 2005. Elementary School 2 section, Living Beings. Available at: http://revistaescola.abril.com.br/ciencias/pratica-pedagogica/como-ensinar-microbiologia- 426117.shtml. Accessed on: 02 November 2011.

GIANI, K. **Experimentation in science teaching: possibilities and limits in the search for meaningful learning.** 2006. Master's dissertation. University of Brasilia. Brasília, 2010.

GIANNOTTI, Sirlene. **To give form is to be formed: creative processes in art and childhood.** Master's thesis. São Paulo, FE/USP, 2008.

GIOPPO, Christiane; SCHEFFER, Elizabeth W. o; NEVES, Marcos C. d. o ensino experimental na escola fundamental: uma reflexão no caso do paraná. In: **Educar: revista da editora da UFPR,** Paraná, n.14, p. 39-57,1998.

GOMES, A. D. T. **Recognising and using experimental tests in the school laboratory.** 2005. Dissertation (Master's in Education) - Faculty of Education, UFMG, Belo Horizonte, 2005.

HAMBURGER & LIMA. **The act of teaching science.** In the open. Brasília: INEP, 7 (40): 13-16, Oct/Dec 1988.

HOERNIG, A.M.; PEREIRA A.B. As aulas de Ciências Iniciando pela Prática: O que Pensam os Alunos.

REVISTA DA ASSOCIAÇÃO BRASILEIRA DE PESQUISA EM EDUCAÇÃO EM CIÊNAS, v.4, n.3. 2004, p. 19-28.

LEAO, D. M. M. Contemporary Paradigms of Education: Traditional School and Constructivist School. **CADERNOS DE PESQUISA,** n. 107, p. 187-206, July/1999. Available at: <http://www.scielo.br/pdf/cp/nl07/nl07a08.pdf>. Accessed on: 21 February 2012.

LUNETTA, V. N. 1991. Practical activities in science teaching. **REVISTA PORTUGUESA DE EDUCAÇÃO,** v. 2, n. 1, p. 81-90.

MALACARNE, V.; STRIEDER, D. M. The unveiling of science in the early years of primary education: a look through the lens of experimentation. **VIVÊNCIAS.** Vol.5, N.7:, May/2009, p.75-85. Available at: http://www.reitoria.uri.br/~vivencias/Numero_007/artigos/artigos_vivencias_07/Artigo_10.pd f> Accessed on: 21 February 2012.

MELO, J.F.R. Development of practical experimental activities in the teaching of Biology - A case study. Dissertation (Professional Master's in Science Teaching) - University of Brasília - Brasília. 2010.

MOREIRA, M. A. (1999). **Meaningful learning.** Brasília: Editora Universidade de Brasília
MORI,L.; MIYAKI, C.Y.; ARIAS, M.C. The finches of the Galapagos: what Darwin didn't see, but the Grants did. **GENETICS AT SCHOOL.** 1(1): 1-3. Available at: http://www.geneticanaescola.com.br/Anolvoll.html. Accessed on 04 January 2013.

PEDRANCINI, V. D.; CORAZZA-NUNES, M. J.; GALUCH, M. T. B.; MOREIRA, A. L. O.R.; RIBEIRO, A. C. Teaching and learning biology in secondary schools and the appropriation of scientific and biotechnological knowledge. **REVISTA ELECTRÓNICA DE ENSENANZA DE LÃS CIÊNCIAS.** Vol. 6, n. 2, p. 299-309, 2007.

POSSOBOM, C. C. F.; OKADA, F. K.; DINIZ, R. E. S.. **Practical laboratory activities in biology and science teaching: an experience report.** In: Universidade Estadual Paulista - Pró- Reitoria de Graduação. (Org.). Núcleos de Ensino. São Paulo: Editora da UNESP, 2003, v. 1, p. 113-123

PRAIA, J.; CACHAPUZ, A.; GIL-PÉREZ, D. The hypothesis and the scientific experience in science education: contributions to an epistemological reorientation. **CIÊNCIA & EDUCAÇÃO,** v. 8, n.2, p.253-262, 2002.

PRIGOL, S.; GIANNOTTI, S.M. A importância da utilização de práticas no processo de ensino-aprendizagem de ciências naturais enfocando a morfologia da flor. **SIMPÓSIO NACIONAL DE EDUCAÇÃO - XX** Semana da Pedagogia, 2008.

RODRIGUES, L. P., MOURA, L.S., TESTA, E., The traditional and the modern regarding didactics in higher education. **REVISTA CIENTÍFICA DO ITPAC,** Araguaína, v.4, n.3, Pub.5, July 2011. Available at: http://www.itpac.br/hotsite/revista/artigos/43/5.pdf. Accessed on: 23 November 2012

ROSA, C. W. DA; ROSA, A. B. DA- The teaching of science (Physics) in Brazil: from history to new educational guidelines; **REVISTA IBERO AMERICANA DE EDUCACIÓN / REVISTA IBERO-AMERICANA DE EDUCAÇÃO** - n.° 58/2 - 2012.

SÁ, E. **F. Teachers' discourses on science teaching by enquiry.** 2009. Thesis (Doctorate). Federal University of Minas Gerais. Belo Horizonte, 2009.

SEREIA, D. A. de O.; PIRANHA, M. M. Aulas **práticas investigativas: um a experiência no ensino fundamental para a formação de alunos participativos.** State Secretariat of Education of the State of Paraná - SEED/PR, 2010. Available at: <http://www.diaadiaeducacao.pr.gov.br/diaadia/diadia/arquivos/File/conteudo/artigos_teses/Ciencias/Artigos/aulas_prat_investig.pdf>. Accessed on: 15 January 2012.

SILVA. D. J. A. **Experimental chemistry lessons: adapting the textbook to the reality of secondary school students.** 1999. Available at: <http://www.sbq.org.br/ranteriores/23/resumos/0269-l/index.html.> Accessed on: 25 January 2013

SILVA, D. P. **Issues proposed in the planning of experimental activities of an investigative nature in chemistry teaching: reflections from a group of teachers. 2011.**
Master's dissertation. University of São Paulo. São Paulo, 2011.

SILVA JÚNIOR, A. N. da; BARBOSA, J. R. A. Rethinking the Teaching of Science and Biology in Basic Education: the Path to the Construction of Scientific and Biotechnological Knowledge. **DEMOCRATISE,** v. 3. n.l, p. 1-15 , jan./abr. 2008.

SILVA, R.P; ARAÚJO, M.L.F. Conceptions of experimental activities and implications for the teaching practice of science teachers. **V INTERNATIONAL COLLOQUIUM - EDUCATION AND CONTEMPORANEITY,** 2011. Available at: http://www.educonufs.com.br/vcoloquio/cdcoloquio/cdroom/eixo%206/PDF/Microsoft%20W ord%20-%20CONCEPCOES%20OF%20EXPERIMENTAL%20ACTIVITIES.pdf. Accessed on: 14 November 2012.

SOUZA S.S.P. **Atividades Investigativas como estratégia para o ensino-aprendizagem em ciências: Proposals and learning.** 2007. Master's dissertation. Federal University of Pará. Belém, 2007.

SUART, R. C.; MARCONDES, M. E. R.; CARMO, M. P. Experimental Activities: Using the Energy Involved in Chemical Reactions to Develop Cognitive Skills. **VII ENCONTRO NACIONAL DE PESQUISA EM EDUCAÇÃO EM CIÊNCIAS,** 2009.

TAMIR, P. How are the laboratories used? **JOURNAL OF RESEARCH IN SCIENCE TEACHING,** 14 (4), p. 311-316, 1977.

TAPIA, J. A.; FITA, E. C. **Motivation in the classroom: what is it? How to do it.** São Paulo: Loyola, 2004.

VASCONCELOS, A. L.S.; COSTA, C.H.C.; SANTANA, J.R.; CECCATTO, V. M.; **Importance of the practical approach in teaching Biology for teacher training** (Full degree in Science/Habilitation in Biology/Chemistry - UECE) in Limoeiro do Norte - CE. Fortaleza - CE, 2002. Available at: <http://www.multimeios.ufc.br/arquivos/pc/congressos/congressos-importancia-da- abordagem-pratica-no-ensino-de-biologia.pdf>. Accessed on: 08 September 2012.

ZÔMPERO, A. F.; LABURÚ, C. E. Investigative activities in science teaching: Historical aspects and different approaches, **REVISTA. ENSAIO,** Belo Horizonte, v.13, n.03, p.67-80, Sep-Dec, 2011.

ZULIANI, S. R.Q.A. **Chemistry teaching practice and investigative methodology: A phenomenological reading based on social semiotics.** 2006. Thesis (Doctorate). Federal University of São Carlos. São Carlos, 2006.

APENDICES

APPENDIX A

PROJECT: New directions for experimental lessons in science teaching: Myths, Mysteries and Needs.

Student____

QUIZ - MICROORGANISMS

1. Where are microorganisms found? Tick the correct box.

( ) Soil

( )Air

( ) Man

( ) Food

( ) All alternatives are correct

2. In relation to contact with humans, microorganisms occur:

a) ( ) in a positive way (Indispensable to life)
b) ( ) negatively (harmful effects on health)
c) ( ) neutrally
d) ( ) positively and negatively

3. Are all microorganisms apparently the same?

( ) yes ( ) no

4. Microorganisms can only be seen with the use of (a):

( ) Magnifying glass

( ) Microscope

( ) Eye

( ) Lunette

5. understands microorganisms:

( ) Fungi, bacteria and viruses.

( ) Fungi, viruses and protozoa.

( ) Fungi, bacteria, viruses, protozoa and unicellular algae.

( ) animals, bacteria and protozoa.

6. The use of microorganisms to clean up environmental areas contaminated by pollutants is a simple strategy that is less expensive and has less impact on the environment. This type of action is carried out using bacteria and is known as:

a) Bioremediation

b) Antibiotics

c) Landfill

d) Arqueas

e) Environmental vaccine

7. Tick V for true and F for false.

( ) All the cells that make up our intestines are made up of bacteria.

( ) Some medicines are produced from microorganisms.

( ) All microorganisms cause some kind of disease.

( ) Microorganisms are used in the production of beverages and food.

8. How important are these beings to us?

9. What do microorganisms feed on?

10. Despite being tiny beings, how do we know that microorganisms exist?

11. These are important functions of microorganisms in the environment:

() Degrade some compounds.

( ) It causes pollution.

( ) It has harmful effects on health.

( ) Control of pollutants by biological processes.

( ) Production of oxygen.

12. What measures should be taken to ensure good hygiene?

13. What is needed for microorganisms to proliferate?

( ) Temperature and cleanliness.

( ) Humidity and extreme temperatures.

( ) Water and heat.

( ) A nourishing medium and a favourable environment.

14. Do all cleaning materials act equally on microorganisms? Explain your answer.

APPENDIX B

PROJECT: New directions for experimental lessons in science teaching: Myths, Mysteries and Needs.

Student____

QUESTIONNAIRE - SEEDS

1 . What is a seed?

2 . The seeds are dispensed onto the surface of the globe through:

a) ( ) Wind, water, animals, birds, the fruit itself and man.

b) ( ) Wind, leaves, soil and man

c) ( ) Water, animals, plants and the fruit itself.

d) ( ) Animals, birds, water and leaves.

3 Which seeds do you know? Describe their characteristics.

4 It is correct to say that germination is..:

a) ( ) It is a vegetation layer that lies above the ground.

b) ( ) Seeds are dispersed by the wind.

c) ( ) It is the initial process of plant growth.

d) ( ) It is the reserve of nutrients present in the seed.

5. Tick the extrinsic conditions (environmental conditions) necessary for seed germination.

( ) maturity

( ) water

( )air

( ) Good constitution

( ) Light

6.0 What is inside seeds that makes them grow into plants?

7. These are the functions of the seed:

( ) Provide nutrients that ensure the initial development of germination.

( ) Transport of nutrients throughout the plant.

( ) Expending energy to produce nutrients.

( ) To give protection to the embryo.

8. What elements do plants need to grow?

9. Tick the boxes that refer to the functions of the root:

( ) seed germination

( ) assimilation and transport of water and nutrients

( ) Fixation of the plant to the substrate.

( ) Photosynthesis.

10. How important were seeds in the adaptation of plants to the terrestrial environment?

APPENDIX C

PROJECT: New directions for experimental lessons in science teaching: Myths, Mysteries and Needs.

THE FINCHES OF THE GALAPAGOS

UNDERSTANDING THE ACTIVITY

1. What does each of the tools used to collect seeds represent?

2. What does the seed tray represent?

3. What does each round of seed collection represent?

4. What does the elimination of individuals who fail to collect seeds in a round represent?

5. Set up a table to record the number and variety of seeds that each beak managed to catch.

APPENDIX D
EXPERIMENTAL CLASSROOM

The finches of the Galapagos

OBJECTIVE: To simulate what happens in nature with regard to the availability of food and the morphological characteristics that make it possible to utilise these resources.

INTRODUCTION:

When Charles Darwin passed through the Galapagos archipelago in 1835 on his journey around the world, he was beginning to gather arguments in favour of his theory of the evolution of species.

According to this theory, current animal species are the result of modifications to previous species; the same ancestor can have entirely different descendants. Darwin was to find a perfect example of evolution in the Galapagos, a group of small Pacific islands about 1,000 kilometres off the coast of South America: the finch family that today bears his name. Originating from the same vanished American ancestor, these birds found islands without birds on their arrival.

There were possibilities for all sorts of adaptations, for all sorts of ways of life; and little by little, the Galapagos finches gave rise to 13 different species, which eat different things, which have different beaks. From the huge beak of the grain eater to the pointed beak of the insect eater.

The climate of the Galapagos fluctuates greatly, as does the quantity and variety of fruits and seeds that are the finches' main food. In general, larger individuals are favoured in times of drought (when there are only very hard seeds), while smaller individuals are favoured in very humid conditions (when there is an abundance of seed varieties).

The climate pattern in the Galapagos is characterised by a hot and humid season from January to May, followed by a dry and cooler season. After the rains, the variety of seeds is great, from soft to hard. When there are plenty of seeds, all the birds feed mainly on the softer seeds, which are easier to eat and finish first.

MATERIAL

- Plastic pot containing various seeds;
- Transparent plastic tray;
- Pointless scissors;
- Nail pliers;
- Eyebrow tweezers;
- Clothes peg.

PROCEDURE

1. Distribute a tray containing the seeds to each group of five students;
2. Each student must choose an instrument of a different shape and size;
3. Collect the seeds for 30 seconds using the instruments and not your hands. Each student must keep their seeds separate, and must keep the seeds collected in each of the 30-second rounds separate.
4. After five 30-second collection rounds, check which seeds have been collected and which are left over;
5. If a student doesn't manage to pick up any seeds, they won't be able to take part in the next round;

APPENDIX E

EXPERIMENTAL CLASSROOM
Cultivating microorganisms

OBJECTIVE: To show the existence of microbes and how they contaminate the culture medium.

INTRODUCTION:

Microorganisms exist everywhere, from the earth's surface to high altitudes, in all continental areas, in marine environments from the surface to the abyssal depths. Animals carry huge microbial populations, for example, the human body has 10 billion cells and 100 billion microorganisms.

Of the thousands of known microorganisms, only a small number cause disease. All animals and plants depend on chemical transformations carried out by microorganisms in the environment. Microbes promote the recycling of matter in nature, transforming complex compounds into simpler ones. The degradation products are absorbed by plants, and the plants are then ingested by animals. Finally, plants, animals and their waste are deposited in the environment and the process repeats itself. In the absence of microorganisms, organic matter would accumulate indefinitely.

Microorganisms are made up of different groups of beings: Bacteria, Protozoa, Unicellular algae, fungi and viruses. **Bacteria** - unicellular organisms, i.e. made up of a single cell, are responsible for decomposing organic matter spontaneously in nature and cause a series of diseases in a wide variety of organisms.
Protozoa - also unicellular, they are microorganisms linked to various diseases.

Algae - photosynthesising eukaryotic organisms that play a fundamental role in the balance of nature, as they are extremely abundant autotrophic organisms and are responsible for 30% to 50% of the oxygen produced on the planet.

Fungi - a group of eukaryotic organisms that are larger than bacteria and form a visible mass, those cotton-like branches that can be seen on decaying food. Technologically, man has used these microorganisms for centuries to produce food and more recently to produce medicines.
Viruses - They are completely different from other microorganisms in that they are nothing more than genetic material covered by a protein coat and an envelope, and they need a host cell to reproduce. They cause countless diseases in all known living organisms.
RESOURCES NEEDED: colourless gelatine or agar, meat broth, water, cotton buds, plastic wrap, labels, pens and Petri dishes.

EXPERIMENT PROCEDURE: Divided into groups, the students have to swab places where they believe there are microorganisms. The swab is rubbed lightly over the culture medium to contaminate it. Cover the Petri dishes. Mark on the adhesive labels what type of contamination has taken place. After three days, observe the changes.

APPENDIX F
EXPERIMENTAL LESSON SCRIPT
TESTING CLEANING PRODUCTS

OBJECTIVE: to prove the effectiveness of disinfectants and other products that promise to kill micro-organisms.

RESOURCES NEEDED: bacteria from the first experiment, Petri dish with culture medium, filter paper, tweezers, test tube, disinfectant, bleach and oven.

PROCEDURE: Scrape off some of the bacteria from the already contaminated plates, dilute them in a few drops of water (use a test tube) and spread the mixture of water and bacteria in the Petri dish with the culture medium. Using tweezers, dip the paper filter in disinfectant (if using the bacteria created by dirt on the floor, finger or paper bill) or antiseptic mouthwash (if using those originating from dental plaque). Place the filter in the middle of the plaque contaminated by bacteria and store it in the oven. Wait a few days. The better the product, the larger the transparent halo that will appear around the paper; if it's bad, nothing will happen.

APPENDIX G
SCRIPT EXPERIMENTAL LESSON
SEED OR NOT?

OBJECTIVE: To determine what a seed is

INTRODUCTION:

The seed is the formation of higher plants, responsible for the perpetuation of species. In other words, the seed is the fertilised and developed ovule.

Parts of the seed:

Seed **coat:** this is the seed's protective covering.
Kernel: the main part of the seed. It corresponds to the nucellus of the ovule, somewhat modified after fertilisation. It is protected by the integument and generally consists of two parts: the embryo and the albumen.
Embryo: this is the main part of the seed. The embryo is responsible for the origin of the new plant.
Albumen: is the food reserve accumulated in the seed.

Spreading the seeds

By spreading, we mean the dispersal of seeds across the surface of the globe.
Many factors favour or directly participate in this phenomenon and are called spreading agents: **Wind, water, animals, birds, the fruit itself and man.**

Seed germination

Germination is the set of phenomena that take place in the seed to give rise to a new plant. This process results in the formation of a new plant of the same species. Seed germination is affected by factors that are divided into two groups: external factors and internal factors.

External or extrinsic factors

These depend on the environment in which the seed is placed. The most important are humidity, oxygen and temperature.

* **humidity, oxygen, temperature.**

Internal or intrinsic factors

These are the ones that depend solely on the seed:

* **the seed must be alive, the seed must be complete, the seed must be ripe.**

When extrinsic and intrinsic conditions are met, the seed germinates.

MATERIAL

- Seeds, sand and fertiliser, peti bottle.

APPENDIX H

INFORMED CONSENT FORM

You are being invited to take part in a research study as a volunteer. If you agree to take part in the study, please sign at the end of this document, which is in two copies. One copy is yours and the other belongs to the researcher responsible.

We guarantee the confidentiality of your information. If you refuse, you will not be penalised in any way.

INFORMATION ABOUT THE RESEARCH:

Project title: New directions for experimental classes in science teaching: Myth,

Mystery and Need.

Principal Investigator: Emily Karoliny da Silva Cunha Souto

Contact telephone number: (83) 9802-3442

This work aims to evaluate the application of experimental lessons as a pedagogical resource for the teaching-learning process in the 7th year of primary school at the Papa Paulo VI Municipal Primary School.

The data for the study will be collected through questionnaires and observations. This material will then be analysed, guaranteeing absolute confidentiality with regard to the questions answered, and the names of the participants will be kept confidential. The work will be disseminated for academic purposes, in the hope that it will contribute to greater knowledge of the subject studied. Participants have the right to withdraw from the study at any time, without any prejudice.

Photographs will be taken to record the dynamics of the activity, making it easier to analyse later.

The data collected will be used in the course completion work (TCC) of the undergraduate Biological Sciences course (CB) at the Centre for Education and Health (CES) of the Federal University of Campina Grande.

The material and interpersonal contact will offer minimal risks to employees and the institution.

Any questions you may have now or at any time can be clarified by contacting the above-mentioned telephone number.

INFORMED CONSENT FORM

By this instrument, which fulfils the legal requirements, you, the research subject, after reading the above information, aware of the services and procedures to which you will be subjected, and having no doubts about what has been read and explained, sign your FREE AND INFORMED CONSENT to participate in the proposed research. It is made clear that the research subject or their legal representative may, at any time, withdraw their FREE AND INFORMED CONSENT and cease to participate in the research study and is aware that all work carried out becomes confidential information, kept under professional secrecy.

New Forest of

Signature of the subject or their legal representative

APPENDIX I - Illustrations

Figure 1: Explanation of the project to one of the participating classes

Figure 2: Resolution of the questionnaires

Figure 3: Experience books

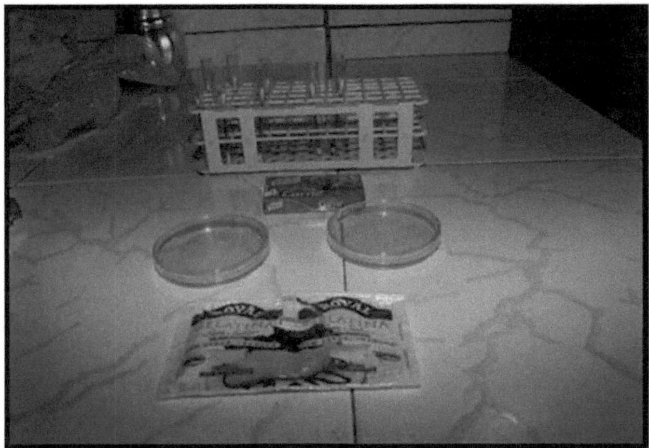

Figure 4: Some of the materials used in the first Microorganisms meeting.

Figure 5: Materials used at the 2° meeting of Microorganisms

Figure 6: Greenhouse for storing culture media

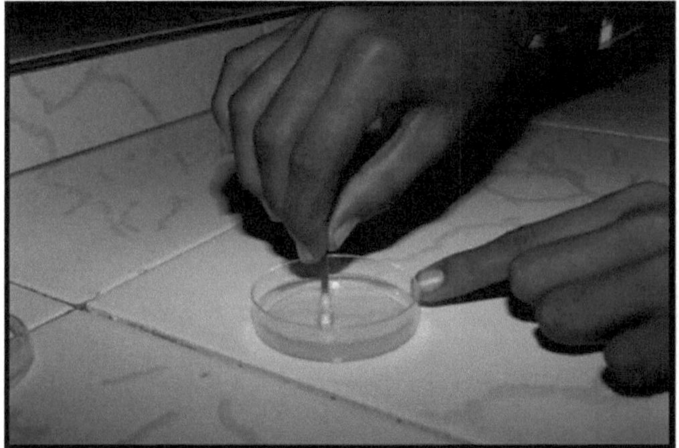

Figure 7. Contamination of the culture medium

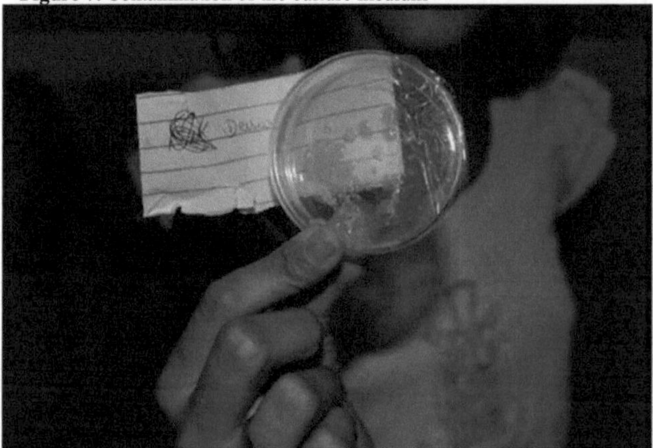

Figure 8. Plate with cultivated microorganisms

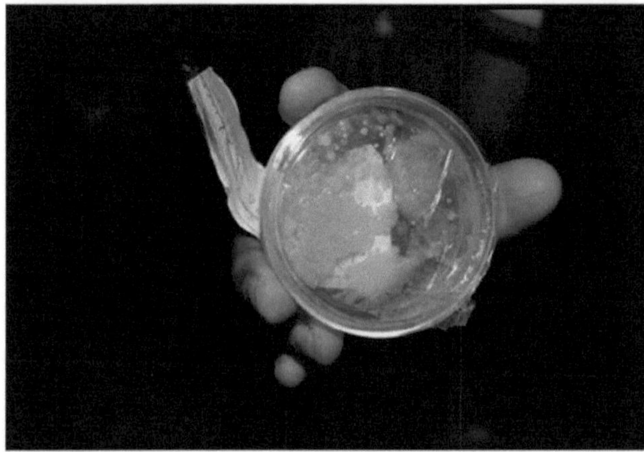

Figure 9. Plate with cultivated microorganisms

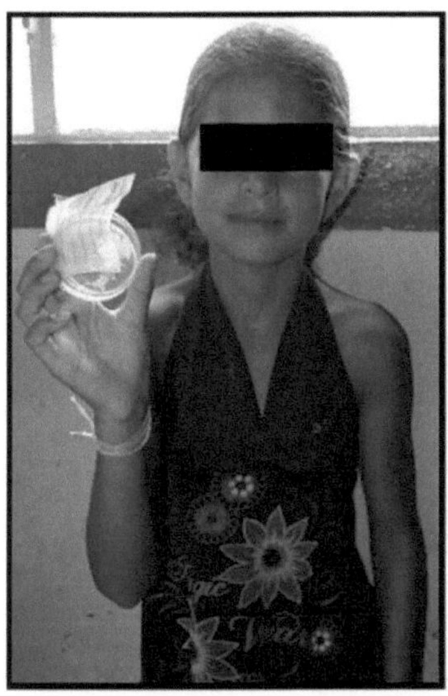

Figure 10. Representative of the group with contaminated culture medium

Figure 11. Instruments used in the experiment on Natural Selection

Figure 12. Instruments used by each group in the Natural Selection experiment

Figure 13. Seed cultivation, I° meeting on the theme of seeds.

Figure 14. Germination of seeds sown at the I° meeting

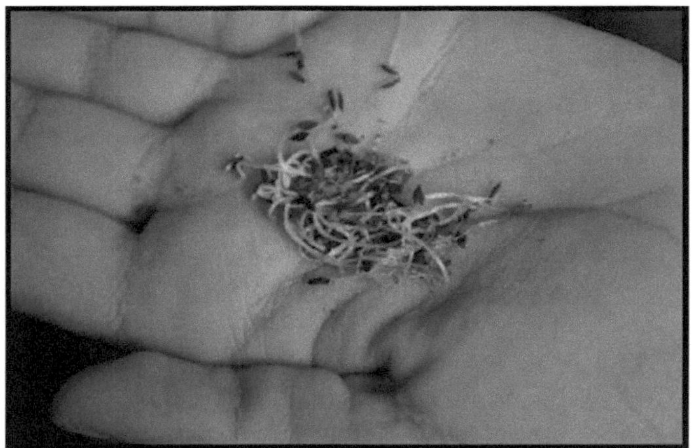

Figure 15. Lettuce seed germination.

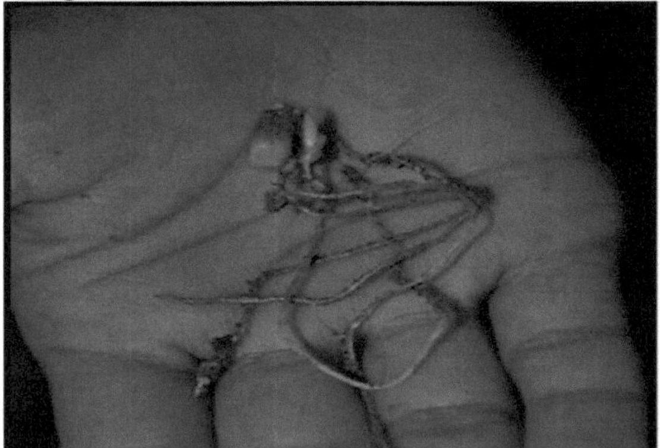

Figure 16. Maize seed germination.

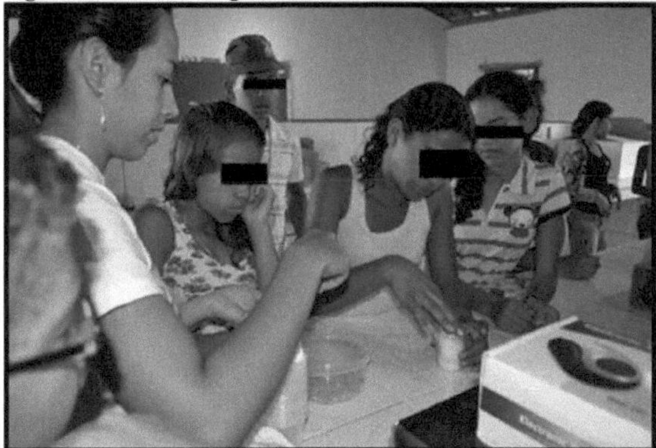

Figure 17. Seed cultivation in cotton.

Figure 18. Seeds grown on cotton to observe the conditions necessary for seed germination.

Printed by Books on Demand GmbH, Norderstedt / Germany